Acoustics: Principles and Applications

Acoustics: Principles and Applications

Martin Hoover

NY RESEARCH
PRESS

New York

Published by NY Research Press
118-35 Queens Blvd., Suite 400,
Forest Hills, NY 11375, USA
www.nyresearchpress.com

Acoustics: Principles and Applications
Martin Hoover

International Standard Book Number: 978-1-63238-889-6 (Hardback)

Cataloging-in-Publication Data

Acoustics : principles and applications / Martin Hoover.
    p. cm.
Includes bibliographical references and index.
ISBN 978-1-63238-889-6
1. Sound. 2. Hearing. 3. Physics. 4. Acoustical engineering. I. Hoover, Martin.
QC225.15 .A26 2022
534--dc23

Table of Contents

This book is a culmination of my many years of practice in this field. I attribute the success of this book to my support group. I would like to thank my parents who have showered me with unconditional love and support and my peers and professors for their constant guidance.

The branch of physics that focuses on the study of mechanical waves in gases, liquids and solids is referred to as acoustics. It includes topics such as vibration, sound, infrasound and ultrasound. Acoustics is used in numerous industries such as audio and noise control industries. It is also used in various other sectors such as medicine, architecture, industrial production and warfare. A few of the major sub-disciplines of acoustics are archaeoacoustics, architectural acoustics, aeroacoustics and psychoacoustics. Archaeoacoustics involves the testing of the acoustic properties of prehistoric sites. Architectural acoustics seeks to control and regulate sound within a building. Aeroacoustics studies the noise produced by air movement and psychoacoustics deals with the perception of sounds by humans. This textbook provides comprehensive insights into the field of acoustics. Some of the diverse topics covered herein address the varied branches that fall under this category. This book will provide comprehensive knowledge to the readers.

The details of chapters are provided below for a progressive learning:

Chapter – Sound and Acoustics

The branch of physics which is concerned with the study of mechanical waves in liquids, gases and solids is referred to as acoustics. Some of the topics which are studied in acoustics are sound, vibration, pitch, wave propagation, etc. This chapter proves a brief introduction to these fundamental concepts of acoustics.

Chapter – Sub-disciplines of Acoustics

Acoustics can be divided into various sub-disciplines. These include aeroacoustics, archaeoacoustics, acoustic signal processing, architectural acoustics, bioacoustics, musical acoustics, underwater acoustics and psychoacoustics. The topics elaborated in this chapter will help in gaining a better perspective about these sub-disciplines of acoustics.

Chapter – Key Concepts in Acoustics

Some of the key concepts of acoustics are acoustic attenuation, acoustic impedance, acoustic absorption, soundproofing and mode conversion. This chapter closely examines these concepts of acoustics to provide a thorough understanding of the subject.

Chapter – Characteristics of Sound

The vibration which typically propagates as an audible wave of pressure through a transmission medium is known as sound. Some of its common characteristics are amplitude, wavelength, timbre, frequency, energy, power and intensity. All these characteristics of sound have been carefully analyzed in this chapter.

Chapter – Categorization of Sound on the basis of Frequency

On the basis of its frequency, sound can be broadly categorized into ultrasound, infrasound and audible sound. Ultrasound has a frequency higher while infrasound has a frequency lower than the limit of human hearing. This chapter closely examines these key categories of sound to provide an extensive understanding of the subject.

Chapter – Units of Sound

The acoustic units of sound measurement are known as sound units. The major units of sound include decibel, phon, sone and hertz. The topics elaborated in this chapter will help in gaining a better perspective about these units of sound.

Chapter – Sound Propagation: Reflection, Refraction and Diffraction

Sound is a vibration which needs a transmission medium for its propagation. During propagation, sound waves can undergo refraction, reflection or diffraction. This chapter closely examines these key concepts of sound propagation to provide an extensive understanding of the subject.

Chapter – Acoustic Transmission

The transmission of sounds through and between materials, such as walls, air and musical instruments, is referred to as acoustic transmission. Some of the major concepts related to acoustic transmission are sound transmission class and sound transmission loss. This chapter has been carefully written to provide an easy understanding of these facets of acoustic transmission.

Chapter – Acoustic Measurement

Acoustic measurement refers to the measurement of the values which describe sound in terms of their intensities as well as qualitative features. Some of its applications include acoustic microscopy, beamforming, spectrograms, and acoustic interferometers. This chapter closely examines these concepts and applications of acoustic measurement to provide an extensive understanding of the subject.

Martin Hoover

1
Sound and Acoustics

The branch of physics which is concerned with the study of mechanical waves in liquids, gases and solids is referred to as acoustics. Some of the topics which are studied in acoustics are sound, vibration, pitch, wave propagation, etc. This chapter proves a brief introduction to these fundamental concepts of acoustics.

Acoustics is the science concerned with the production, control, transmission, reception, and effects of sound.

Beginning with its origins in the study of mechanical vibrations and the radiation of these vibrations through mechanical waves, acoustics has had important applications in almost every area of life. It has been fundamental to many developments in the arts—some of which, especially in the area of musical scales and instruments, took place after long experimentation by artists and were only much later explained as theory by scientists. For example, much of what is now known about architectural acoustics was actually learned by trial and error over centuries of experience and was only recently formalized into a science.

Other applications of acoustic technology are in the study of geologic, atmospheric, and underwater phenomena. Psychoacoustics, the study of the physical effects of sound on biological systems, has been of interest since Pythagoras first heard the sounds of vibrating strings and of hammers hitting anvils in the 6th century BC, but the application of modern ultrasonic technology has only recently provided some of the most exciting developments in medicine. Even today, research continues into many aspects of the fundamental physical processes involved in waves and sound and into possible applications of these processes in modern life. Sound waves follow physical principles that can be applied to the study of all waves.

The origin of the science of acoustics is generally attributed to the Greek philosopher Pythagoras (6th century BC), whose experiments on the properties of vibrating strings that produce pleasing musical intervals were of such merit that they led to a tuning system that bears his name. Aristotle (4th century BC) correctly suggested that a sound wave propagates in air through motion of the air—a hypothesis based more on philosophy than on experimental physics; however, he also incorrectly suggested that high frequencies propagate faster than low frequencies—an error that persisted for many centuries. Vitruvius, a Roman architectural engineer of the 1st century BC, determined the correct mechanism for the transmission of sound waves, and he contributed substantially to the acoustic design of theatres. In the 6th century AD, the Roman philosopher Boethius documented several ideas relating science to music, including a suggestion that the human perception of pitch is related to the physical property of frequency.

The modern study of waves and acoustics is said to have originated with Galileo Galilei, who elevated to the level of science the study of vibrations and the correlation between pitch and frequency of the sound source. His interest in sound was inspired in part by his father, who was a mathematician, musician, and composer of some repute. Following Galileo's foundation work, progress in acoustics came relatively rapidly. The French mathematician Marin Mersennestudied the vibration of stretched strings; the results of these studies were summarized in the three Mersenne's laws. Mersenne's Harmonicorum Libri provided the basis for modern musical acoustics. Later in the century Robert Hooke, an English physicist, first produced a sound wave of known frequency, using a rotating cog wheel as a measuring device. Further developed in the 19th century by the French physicist Félix Savart, and now commonly called Savart's disk, this device is often used today for demonstrations during physics lectures. In the late 17th and early 18th centuries, detailed studies of the relationship between frequency and pitch and of waves in stretched strings were carried out by the French physicist Joseph Sauveur, who provided a legacy of acoustic terms used to this day and first suggested the name acoustics for the study of sound.

One of the most interesting controversies in the history of acoustics involves the famous and often misinterpreted "bell-in-vacuum" experiment, which has become a staple of contemporary physics lecture demonstrations. In this experiment the air is pumped out of a jar in which a ringing bell is located; as air is pumped out, the sound of the bell diminishes until it becomes inaudible. As late as the 17th century many philosophers and scientists believed that sound propagated via invisible particles originating at the source of the sound and moving through space to affect the ear of the observer. The concept of sound as a wave directly challenged this view, but it was not established experimentally until the first bell-in-vacuum experiment was performed by Athanasius Kircher, a German scholar, who described it in his book Musurgia Universalis. Even after pumping the air out of the jar, Kircher could still hear the bell, so he concluded incorrectly that air was not required to transmit sound. In fact, Kircher's jar was not entirely free of air, probably because of inadequacy in his vacuum pump. By 1660 the Anglo-Irish scientist Robert Boyle had improved vacuum technology to the point where he could observe sound intensity decreasing virtually to zero as air was pumped out. Boyle then came to the correct conclusion that a medium such as air is required for transmission of sound waves. Although this conclusion is correct, as an explanation for the results of the bell-in-vacuum experiment it is misleading. Even with the mechanical pumps of today, the amount of air remaining in a vacuum jar is more than sufficient to transmit a sound wave. The real reason for a decrease in sound level upon pumping air out of the jar is that the bell is unable to transmit the sound vibrations efficiently to the less dense air remaining, and that air is likewise unable to transmit the sound efficiently to the glass jar. Thus, the real problem is one of an impedance mismatch between the air and the denser solid materials—and not the lack of a medium such as air, as is generally presented in textbooks. Nevertheless, despite the confusion regarding this experiment, it did aid in establishing sound as a wave rather than as particles.

Measuring the Speed of Sound

Once it was recognized that sound is in fact a wave, measurement of the speed of sound became a serious goal. In the 17th century, the French scientist and philosopher Pierre Gassendi made the earliest known attempt at measuring the speed of sound in air. Assuming correctly that the speed of light is effectively infinite compared with the speed of sound, Gassendi measured the time difference between spotting the flash of a gun and hearing its report over a long distance on a still day. Although the value he obtained was too high—about 478.4 metres per second (1,569.6 feet per second)—he correctly concluded that the speed of sound is independent of frequency. In the 1650s, Italian physicists Giovanni Alfonso Borelli and Vincenzo Viviani obtained the much better value of 350 metres per second using the same technique. Their compatriot G.L. Bianconi demonstrated in 1740 that the speed of sound in air increases with temperature. The earliest precise experimental value for the speed of sound, obtained at the Academy of Sciences in Paris in 1738, was 332 metres per second—incredibly close to the presently accepted value, considering the rudimentary nature of the measuring tools of the day. A more recent value for the speed of sound, 331.45 metres per second (1,087.4 feet per second), was obtained in 1942; it was amendedin 1986 to 331.29 metres per second at 0 °C (1,086.9 feet per second at 32 °F).

The speed of sound in water was first measured by Daniel Colladon, a Swiss physicist, in 1826. Strangely enough, his primary interest was not in measuring the speed of sound in water but in calculating water's compressibility—a theoretical relationship between the speed of sound in a material and the material's compressibility having been established previously. Colladon came up with a speed of 1,435 metres per second at 8 °C; the presently accepted value interpolated at that temperature is about 1,439 metres per second.

Two approaches were employed to determine the velocity of sound in solids. In 1808 Jean-Baptiste Biot, a French physicist, conducted direct measurements of the speed of sound in 1,000 metres of iron pipe by comparing it with the speed of sound in air. A better measurement had earlier been carried out by a German, Ernst Florenz Friedrich Chladni, using analysis of the nodal pattern in standing-wave vibrations in long rods.

Modern Advances

Simultaneous with these early studies in acoustics, theoreticians were developing the mathematical theory of waves required for the development of modern physics, including acoustics. In the early 18th century, the English mathematician Brook Taylor developed a mathematical theory of vibrating strings that agreed with previous experimental observations, but he was not able to deal with vibrating systems in general without the proper mathematical base. This was provided by Isaac Newton of England and Gottfried Wilhelm Leibniz of Germany, who, in pursuing other interests, independently developed the theory of calculus, which in turn allowed the derivation of the general wave equation by the French mathematician and scientist Jean Le Rond d'Alembert in the 1740s. The Swiss mathematicians Daniel Bernoulli and Leonhard Euler, as well as the Italian-French mathematician Joseph-Louis Lagrange, further applied the new equations of calculus to waves in strings and in the air. In the 19th century, Siméon-Denis Poisson of France extended these developments to stretched membranes, and the German mathematician Rudolf Friedrich Alfred Clebsch completed Poisson's earlier studies. A German experimental physicist, August Kundt, developed a number of important techniques for investigating properties of sound waves. These included the Kundt's tube.

One of the most important developments in the 19th century involved the theory of vibrating plates. In addition to his work on the speed of sound in metals, Chladni had earlier introduced a technique of observing standing-wave patterns on vibrating plates by sprinkling sand onto the plates—a demonstration commonly used today. Perhaps the most significant step in the theoretical explanation of these vibrations was provided in 1816 by the French mathematician Sophie Germain, whose explanation was of such elegance and sophistication that errors in her treatment of the problem were not recognized until some 35 years later, by the German physicist Gustav Robert Kirchhoff.

The analysis of a complex periodic wave into its spectral components was theoretically established early in the 19th century by Jean-Baptiste-Joseph Fourier of France and is now commonly referred to as the Fourier theorem. The German physicist Georg Simon Ohm first suggested that the ear is sensitive to these spectral components; his idea that the ear is sensitive to the amplitudes but not the phases of the harmonics of a complex tone is known as Ohm's law of hearing (distinguishing it from the more famous Ohm's law of electrical resistance).

Hermann von Helmholtz made substantial contributions to understanding the mechanisms of hearing and to the psychophysics of sound and music. His book On the Sensations of tone as a Physiological Basis for the theory of music is one of the classics of acoustics. In addition, he constructed a set of resonators, covering much of the audio spectrum, which were used in the spectral analysis of musical tones. The Prussian physicist Karl Rudolph Koenig, an extremely clever and creative experimenter, designed many of the instruments used for research in hearing and music, including a frequency standard and the manometric flame. The flame-tube device, used to render standing sound waves "visible," is still one of the most fascinating of physics classroom demonstrations. The English physical scientist John William Strutt, 3rd Baron Rayleigh, carried out an enormous variety of acoustic research; much of it was included in his two-volume treatise, The Theory of Sound, publication of which in 1877–78 is now thought to mark the beginning of modern acoustics.

The study of ultrasonics was initiated by the American scientist John LeConte, who in the 1850s developed a technique for observing the existence of ultrasonic waves with a gas flame. This technique was later used by the British physicist John Tyndall for the detailed study of the properties of sound waves. The piezoelectric effect, a primary means of producing and sensing ultrasonic waves, was discovered by the French physical chemist Pierre Curie and his brother Jacques in 1880. Applications of ultrasonics, however, were not possible until the development in the early 20th century of the electronic oscillator and amplifier, which were used to drive the piezoelectric element.

Among 20th-century innovators were the American physicist Wallace Sabine, considered to be the originator of modern architectural acoustics, and the Hungarian-born American physicist Georg von Békésy, who carried out experimentation on the ear and hearing and validated the commonly accepted place theory of hearing first suggested by Helmholtz. Békésy's book Experiments in Hearing, published in 1960, is the magnum opus of the modern theory of the ear.

Amplifying, Recording and Reproducing

The earliest known attempt to amplify a sound wave was made by Athanasius Kircher, of "bell-in-vacuum" fame; Kircher designed a parabolic horn that could be used either as a hearing aid or as a voice amplifier. The amplification of body sounds became an important goal, and the first stethoscope was invented by a French physician, René Laënnec, in the early 19th century.

Attempts to record and reproduce sound waves originated with the invention in 1857 of a mechanical sound-recording device called the phonautograph by Édouard-Léon Scott de Martinville. The first device that could actually record and play back sounds was developed by the American inventor Thomas Alva Edison in 1877. Edison's phonograph employed grooves of varying depth in a cylindrical sheet of foil, but a spiral groove on a flat rotating disk was introduced a decade later by the German-born American inventor Emil Berliner in an invention he called the gramophone. Much significant progress in recording and reproduction techniques was made during the first half of the 20th century, with the development of high-quality electromechanical transducers and linear electronic circuits. The most important improvement on the standard phonograph record in the second half of the century was the compact disc, which employed digital techniques developed in mid-century that substantially reduced noise and increased the fidelity and durability of the recording.

Architectural Acoustics

Reverberation Time

Although architectural acoustics has been an integral part of the design of structures for at least 2,000 years, the subject was only placed on a firm scientific basis at the beginning of the 20th century by Wallace Sabine. Sabine pointed out that the most important quantity in determining the acoustic suitability of a room for a particular use is its reverberation time, and he provided a scientific basis by which the reverberation time can be determined or predicted.

When a source creates a sound wave in a room or auditorium, observers hear not only the sound wave propagating directly from the source but also the myriadreflections from the walls, floor, and ceiling. These latter form the reflected wave, or reverberant sound. After the source ceases, the reverberant sound can be heard for some time as it grows softer. The time required, after the sound source ceases, for the absolute intensity to drop by a factor of 10^6—or, equivalently, the time for the intensity level to drop by 60 decibels—is defined as the reverberation time (RT, sometimes referred to as RT_{60}). Sabine recognized that the reverberation time of an auditorium is related to the volume of the auditorium and to the ability of the walls, ceiling, floor, and contents of the room to absorb sound. Using these assumptions, he set forth the empirical relationship through which the reverberation time could be determined: $RT = {}^{0.05V}/A$, where RT is the reverberation time in seconds, V is the volume of the room in cubic feet, and A is the total sound absorption of the room, measured by the unit sabin. The sabin is the absorption equivalent to one square foot of perfectly absorbing surface—for example, a one-square-foot hole in a wall or five square feet of surface that absorbs 20 percent of the sound striking it.

Both the design and the analysis of room acoustics begin with this equation. Using the equation and the absorption coefficients of the materials from which the walls are to be constructed, an approximation can be obtained for the way in which the room will function acoustically. Absorbers and reflectors, or some combination of the two, can then be used to modify the reverberation time and its frequencydependence, thereby achieving the most desirable characteristics for specific uses. Representative absorption coefficients—showing the fraction of the wave, as a function of frequency, that is absorbed when a sound hits various materials—are given in the Table. The absorption from all the surfaces in the room are added together to obtain the total absorption (A).

Absorption Coefficients of Common Materials at Several Frequencies						
Material	Frequency (Hertz)					
	125	250	500	1,000	2,000	4,000
Concrete	0.01	0.01	0.02	0.02	0.02	0.03
Plasterboard	0.20	0.15	0.10	0.08	0.04	0.02
Acoustic board	0.25	0.45	0.80	0.90	0.90	0.90
Curtains	0.05	0.12	0.25	0.35	0.40	0.45

While there is no exact value of reverberation time that can be called ideal, there is a range of values deemed to be appropriate for each application. These vary with the size of the room, but the averages can be calculated and indicated by lines on a graph. The need for clarity in understanding speech dictates that rooms used for talking must have a reasonably short reverberation time. On the other hand, the full sound desirable in the performance of music of the Romantic era, such as Wagner operas or Mahler symphonies, requires a long reverberation time. Obtaining a clarity suitable for the light, rapid passages of Bach or Mozart requires an intermediate value of reverberation time. For playing back recordings on an audio system, the reverberation time should be short, so as not to create confusion with the reverberation time of the music in the hall where it was recorded.

Acoustic Criteria

Many of the acoustic characteristics of rooms and auditoriums can be directly attributed to specific physically measurable properties. Because the music critic or performing artist uses a different vocabulary to describe these characteristics than does the physicist, it is helpful to survey some of the more important features of acoustics and correlate the two sets of descriptions.

"Liveness" refers directly to reverberation time. A live room has a long reverberation time and a dead room a short reverberation time. "Intimacy" refers to the feeling that listeners have of being physically close to the performing group. A room is generally judged intimate when the first reverberant sound reaches the listener within about 20 milliseconds of the direct sound. This condition is met easily in a small room, but it can also be achieved in large halls by the use of orchestral shells that partially enclose the performers. Another example is a canopy placed above a speaker in a large room such as a cathedral: this leads to both a strong and a quick first reverberation and thus to a sense of intimacy with the person speaking.

The amplitude of the reverberant sound relative to the direct sound is referred to as fullness. Clarity, the opposite of fullness, is achieved by reducing the amplitude of the reverberant sound. Fullness generally implies a long reverberation time, while clarity implies a shorter reverberation time. A fuller sound is generally required of Romantic music or performances by larger groups, while more clarity would be desirable in the performance of rapid passages from Bach or Mozart or in speech.

"Warmth" and "brilliance" refer to the reverberation time at low frequencies relative to that at higher frequencies. Above about 500 hertz, the reverberation time should be the same for all frequencies. But at low frequencies an increase in the reverberation time creates a warm sound, while, if the reverberation time increased less at low frequencies, the room would be characterized as more brilliant.

"Texture" refers to the time interval between the arrival of the direct sound and the arrival of the first few reverberations. To obtain good texture, it is necessary that the first five reflections arrive at the observer within about 60 milliseconds of the direct sound. An important corollary to this requirement is that the intensity of the reverberations should decrease monotonically; there should be no unusually large late reflections.

"Blend" refers to the mixing of sounds from all the performers and their uniform distribution to the listeners. To achieve proper blend it is often necessary to place a collection of reflectors on the stage that distribute the sound randomly to all points in the audience.

Although the above features of auditorium acoustics apply to listeners, the idea of ensemble applies primarily to performers. In order to perform coherently, members of the ensemble must be able to hear one another. Reverberant sound cannot be heard by the members of an orchestra, for example, if the stage is too wide, has too high a ceiling, or has too much sound absorption on its sides.

Acoustic Problems

Certain acoustic problems often result from improper design or from construction limitations. If large echoes are to be avoided, focusing of the sound wave must be avoided. Smooth, curved reflecting surfaces such as domes and curved walls act as focusing elements, creating large echoes and leading to bad texture. Improper blend results if sound from one part of the ensemble is focused to one section of the audience. In addition, parallel walls in an auditorium reflect sound back and forth, creating a rapid, repetitive pulsing of sound known as flutter echo and even leading to destructive interference of the sound wave. Resonances at certain frequencies should also be avoided by use of oblique walls.

Acoustic shadows, regions in which some frequency regions of sound are attenuated, can be caused by diffraction effects as the sound wave passes around large pillars and corners or underneath a low balcony. Large reflectors called clouds, suspended over the performers, can be of such a size as to reflect certain frequency regions while allowing others to pass, thus affecting the mixture of the sound.

External noise can be a serious problem for halls in urban areas or near airports or highways. One technique often used for avoiding external noise is to construct the auditorium as a smaller room within a larger room. Noise from air blowers or other mechanical vibrations can be reduced using techniques involving impedance and by isolating air handlers.

Good acoustic design must take account of all these possible problems while emphasizing the desired acoustic features. One of the problems in a large auditorium involves simply delivering an adequate amount of sound to the rear of the hall. The intensity of a spherical sound wave decreases in intensity at a rate of six decibels for each factor of two increase in distance from the source. If the auditorium is flat, a hemispherical wave will result. Absorption of the diffracted wave by the floor or audience near the bottom of the hemisphere will result in even greater absorption, so that the resulting intensity level will fall off at twice the theoretical rate, at about 12 decibels for each factor of two in distance. Because of this absorption, the floors of an auditorium are generally sloped upward toward the rear.

Sound

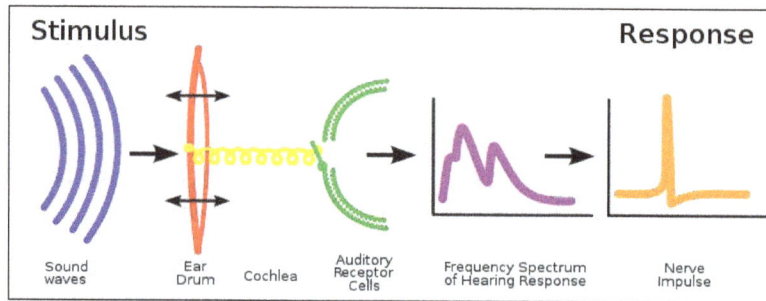

A schematic representation of hearing. (Blue: sound waves. Red: eardrum. Yellow: cochlea. Green: auditory receptor cells. Purple: frequency spectrum of hearing response. Orange: nerve impulse.)

In regular usage, the term sound is applied to any stimulus that excites our sense of hearing. The cause of sound is vibratory movement from a disturbance, communicated to the ear through a medium such as air. Scientists group all such vibratory phenomena under the general category of "sound," even when they lie outside the range of human hearing. The scientific study of sound is known as acoustics.

We depend on sound for communication through speech and artistic expression through music. Our ability to hear sounds provides us with an important mode of perception of our environment. Certain sounds of nature and music have the power to calm the mind and heal the body. Extremely loud noises, on the other hand, can damage our sense of hearing.

Through the development of technology, sound waves can be converted to electrical signals and radio waves and transmitted almost instantaneously to distant parts of the globe. In addition, sounds can be amplified, modified, stored, and replayed at will. Ultrasound (high-frequency sound) is used to generate images of a fetus or a person's internal organs, and to inspect materials for possible flaws. The technology of sonar, based on underwater sound propagation, is useful for detecting icebergs, marine life, and ocean-going vessels.

Properties of Sound

Solids, liquids, and gases are all capable of transmitting sound. For example, the practice of putting one's ear to the ground to listen for an approaching train is based on the fact that solids can transmit sound. Likewise, one can hear sounds when one's head is submerged in a swimming pool, thus demonstrating the ability of a liquid to carry sound. The matter that supports the transmission of sound is called the medium.

Sound is transmitted by means of sound waves, much as a pebble thrown into a lake generates waves on the surface of the water. In air, a sound wave is a disturbance that creates a region of high pressure (compression) followed by one of low pressure (rarefaction). These variations in pressure are transferred to adjacent regions of the air in the form of a spherical wave radiating outward from the disturbance. Sound is therefore characterized by the properties of waves, such as their frequency, wavelength, period, amplitude, and velocity (or speed).

Sound waves are longitudinal waves, meaning that the vibrations (compression and rarefaction of the medium) occur in the direction in which the wave moves. By contrast, the waves on a string are transverse waves, meaning that the vibrations are perpendicular to the direction in which the wave moves.

The properties of a sound wave depend upon the springiness, or elasticity, of the material that the sound travels through. In a gas, stresses and strains are manifested as changes in pressure and density. The movement of a sound wave is accompanied by the transmission of energy that is spread over the spherical wave front.

The term noise is usually applied to an unwanted sound. In science and engineering, noise is an undesirable component that obscures a signal. On the other hand, noises can also be useful at times. For instance, a noisy car engine warns the driver of engine trouble; a noisy infant is calling for attention.

Usefulness in Human Society

Language is communicated primarily though sound. Oral tradition was the earliest means of recording human history. In addition, the oldest artifacts of human civilization include musical instruments such as the flute. The design of musical instruments requires an understanding of the manner in which sound is created and transmitted, and a knowledge of materials and their acoustical characteristics. Certain naturally occurring and artificially produced sounds can soothe or stimulate the mind and help heal the body.

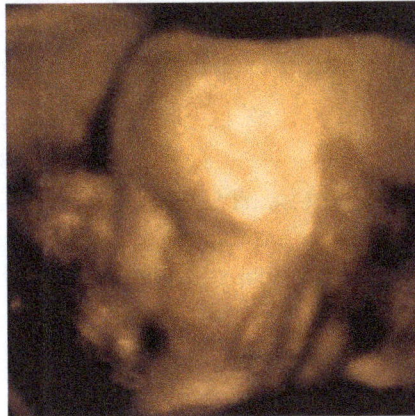

A baby in its mother's womb, at age 29 weeks, was obtained by "3D ultrasound"

Ancient societies constructed amphitheaters designed to carry the sounds of actors' voices to the audience, requiring knowledge of acoustics. Modern performance spaces offer challenges similar to those that faced the ancients. Modern sound technology is closely intertwined with the electronics industry, which has perfected a multitude of ways to convey and reproduce sound electronically.

The telephone, one of the earliest technologies developed for rapid communication, translates sound waves into electrical impulses that are converted back into sound waves at the receiving end. Recording devices store sound wave information, such as conversations or musical performances, by translating them into a mechanical or electronic form that can be used to reconstitute the original sound.

An amplifier takes a sound of weak amplitude and generates an equivalent one with greater amplitude that can be heard more easily. Microphones and sound systems make use of this technology.

Sound can also be used to acquire information about objects at a distance or otherwise hidden from sight. For example, ships and submarines use sonar to detect icebergs, fish, and other objects in the water. Also, a medical imaging technique called sonography uses ultrasound (high-frequency sound waves) to visualize a developing fetus or a patient's internal organs. In industry, ultrasound is a useful means to detect flaws in materials.

Perception of Sound

Humans and many animals use their ears to hear sound, but loud sounds and low-frequency sounds can be perceived by other parts of the body as well, through the sense of touch. The range of frequencies that humans can hear is approximately between 20 and 20,000 hertz (Hz). This range constitutes the audible spectrum, but it varies from one individual to the next and generally shrinks with age, mostly in the upper part of the spectrum. Some people (particularly women) can hear above 20,000 Hz. The ear is most sensitive to frequencies around 3,500 Hz. Sounds above 20,000 Hz are classified as ultrasound; sounds below 20 Hz, as infrasound.

The amplitude of a sound wave is specified in terms of its pressure, measured in pascal (Pa) units. As the human ear can detect sounds with a very wide range of amplitudes, sound pressure is often reported in terms of what is called the sound pressure level (SPL) on a logarithmic decibel (dB) scale.

The quietest sounds that humans can hear have an amplitude of approximately 20 µPa (micropascals), or a sound pressure level of 0 dB re 20 µPa (often incorrectly abbreviated as 0 dB SPL). (When using sound pressure levels, it is important to always quote the reference sound pressure used. Commonly used reference sound pressures are 20 µPa in air and 1 µPa in water.)

Prolonged exposure to a sound pressure level exceeding 85 dB can permanently damage the ear, sometimes resulting in tinnitus and hearing impairment. Sound levels in excess of 130 dB are considered above of what the human ear can withstand and may result in serious pain and permanent damage. At very high amplitudes, sound waves exhibit nonlinear effects, including shock.

Formula for Sound Pressure Level

The mathematical equation to calculate the sound pressure level (L_p) is as follows.

$$L_p = 10 \log_{10} \left(\frac{p^2}{p_0^2} \right) = 20 \log_{10} \left(\frac{p}{p_0} \right) dB$$

where p is the root-mean-square sound pressure and p_0 is the reference sound pressure.

Examples of Sound Pressure and Sound Pressure Levels

Source of Sound	Sound Pressure (pascal)	Sound Pressure Level (dB re 20µPa)
Threshold of pain	100 Pa	134 dB
Hearing damage during short term effect	20 Pa	approx. 120 dB
Jet, 100 m distant	6 – 200 Pa	110 – 140 dB

Jack hammer, 1 m distant / discotheque	2 Pa	approx. 100 dB
Hearing damage during long-term effect	6×10^{-1} Pa	approx. 90 dB
Major road, 10 m distant	$2\times10^{-1} - 6\times10^{-1}$ Pa	80 – 90 dB
Passenger car, 10 m distant	$2\times10^{-2} - 2\times10^{-1}$ Pa	60 – 80 dB
TV set at home level, 1 m distant	2×10^{-2} Pa	ca. 60 dB
Normal talking, 1 m distant	$2\times10^{-3} - 2\times10^{-2}$ Pa	40 – 60 dB
Very calm room	$2\times10^{-4} - 6\times10^{-4}$ Pa	20 – 30 dB
Leaves noise, calm breathing	6×10^{-5} Pa	10 dB
Auditory threshold at 2 kHz	2×10^{-5} Pa	0 dB

Speed of Sound

The speed of sound has been a subject of study since the days of the philosopher Aristotle. In his writings, Aristotle discussed the time lapse between the sighting of an event and detection of the sound it produces. A cannon, for example, will be seen to flash and smoke before the sound of the explosive powder reaches an observer.

The speed at which sound travels depends on the medium through which the sound waves pass, and is often quoted as a fundamental property of the material. The speed of sound in air or a gas increases with the temperature of the gas. In air at room temperature, the speed of sound is approximately 345 meters per second (ms^{-1}); in water, 1,500 m/s^{-1}; and in a bar of steel, 5,000 m/s^{-1}.

Based on the dynamic properties of matter, Isaac Newton derived a mathematical expression for the speed of sound waves in an elastic or compressible medium. For a gas, this expression reduces to:

$v = (P/\rho)^{1/2}$ (where P = pressure; ρ = density of the gas)

This formula, however, yields a number that is short of the true velocity. The formula was improved upon by eighteenth-century mathematician-physicist Pierre-Simon Laplace, who took into consideration the temperature effects of the compression of the air at the front of a sound wave and derived the following equation:

$v = (\gamma P/\rho)^{1/2}$

where γ is a constant that depends on the heat-retaining properties of the gas.

Longitudnal and Transverse Sound Waves

Waves come in many shapes and forms. While all waves share some basic characteristic properties and behaviors, some waves can be distinguished from others based on some observable (and some non-observable) characteristics. It is common to categorize waves based on these distinguishing characteristics.

One way to categorize waves is on the basis of the direction of movement of the individual particles of the medium relative to the direction that the waves travel. Categorizing waves on this basis leads to three notable categories: transverse waves, longitudinal waves, and surface waves.

A transverse wave is a wave in which particles of the medium move in a direction perpendicular to the direction that the wave moves. Suppose that a slinky is stretched out in a horizontal direction across the classroom and that a pulse is introduced into the slinky on the left end by vibrating the first coil up and down. Energy will begin to be transported through the slinky from left to right. As the energy is transported from left to right, the individual coils of the medium will be displaced upwards and downwards. In this case, the particles of the medium move perpendicular to the direction that the pulse moves. This type of wave is a transverse wave. Transverse waves are always characterized by particle motion being perpendicular to wave motion.

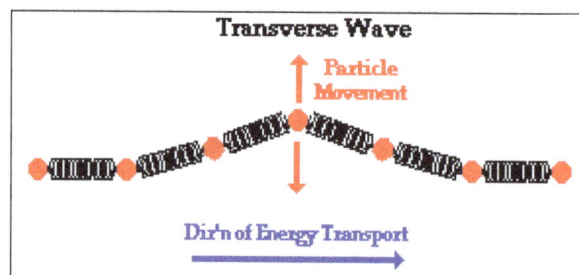

Transverse Wave

Particle Movement

Dir'n of Energy Transport

A longitudinal wave is a wave in which particles of the medium move in a direction parallel to the direction that the wave moves. Suppose that a slinky is stretched out in a horizontal direction across the classroom and that a pulse is introduced into the slinky on the left end by vibrating the first coil left and right. Energy will begin to be transported through the slinky from left to right. As the energy is transported from left to right, the individual coils of the medium will be displaced leftwards and rightwards. In this case, the particles of the medium move parallel to the direction that the pulse moves. This type of wave is a longitudinal wave. Longitudinal waves are always characterized by particle motion being parallel to wave motion.

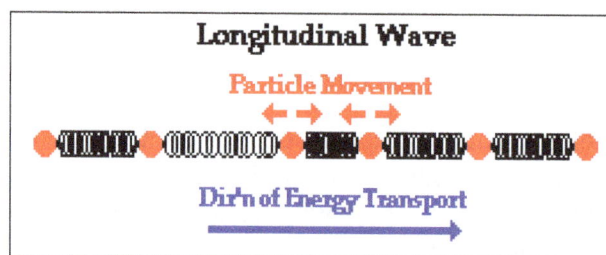

Longitudinal Wave

Particle Movement

Dir'n of Energy Transport

A sound wave traveling through air is a classic example of a longitudinal wave. As a sound wave moves from the lips of a speaker to the ear of a listener, particles of air vibrate back and forth in the same direction and the opposite direction of energy transport. Each individual particle pushes on its neighboring particle so as to push it forward. The collision of particle #1 with its neighbor serves to restore particle #1 to its original position and displace particle #2 in a forward direction. This back and forth motion of particles in the direction of energy transport creates regions within the medium where the particles are pressed together and other regions where the particles are spread apart. Longitudinal waves can always be quickly identified by the presence of such regions. This process continues along the chain of particles until the sound wave reaches the ear of the listener.

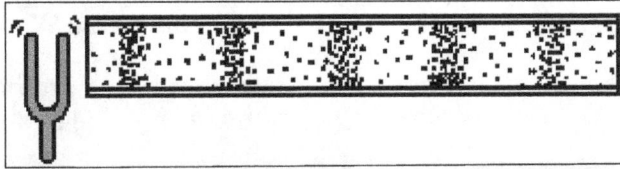

A vibrating tuning fork will force air within a pipe to begin vibrating back and forth
in a direction parallel to the energy transport; sound is a longitudinal wave.

Waves traveling through a solid medium can be either transverse waves or longitudinal waves. Yet waves traveling through the bulk of a fluid (such as a liquid or a gas) are always longitudinal waves. Transverse waves require a relatively rigid medium in order to transmit their energy. As one particle begins to move it must be able to exert a pull on its nearest neighbor. If the medium is not rigid as is the case with fluids, the particles will slide past each other. This sliding action that is characteristic of liquids and gases prevents one particle from displacing its neighbor in a direction perpendicular to the energy transport. It is for this reason that only longitudinal waves are observed moving through the bulk of liquids such as our oceans. Earthquakes are capable of producing both transverse and longitudinal waves that travel through the solid structures of the Earth. When seismologists began to study earthquake waves they noticed that only longitudinal waves were capable of traveling through the core of the Earth. For this reason, geologists believe that the Earth's core consists of a liquid - most likely molten iron.

While waves that travel within the depths of the ocean are longitudinal waves, the waves that travel along the surface of the oceans are referred to as surface waves. A surface wave is a wave in which particles of the medium undergo a circular motion. Surface waves are neither longitudinal nor transverse. In longitudinal and transverse waves, all the particles in the entire bulk of the medium move in a parallel and a perpendicular direction (respectively) relative to the direction of energy transport. In a surface wave, it is only the particles at the surface of the medium that undergo the circular motion. The motion of particles tends to decrease as one proceeds further from the surface.

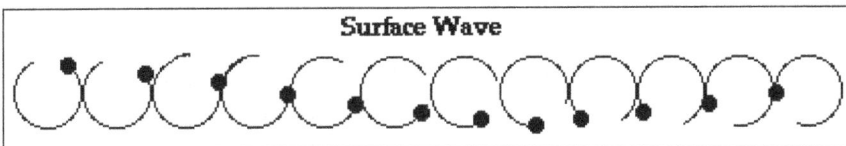

Surface Wave

A surface wave is sometime referred to as a circular wave since particles of
the medium undergo a motion in a complete circle.

Any wave moving through a medium has a source. Somewhere along the medium, there was an initial displacement of one of the particles. For a slinky wave, it is usually the first coil that becomes displaced by the hand of a person. For a sound wave, it is usually the vibration of the vocal chords or a guitar string that sets the first particle of air in vibrational motion. At the location where the wave is introduced into the medium, the particles that are displaced from their equilibrium position always moves in the same direction as the source of the vibration. So if you wish to create a transverse wave in a slinky, then the first coil of the slinky must be displaced in a direction perpendicular to the entire slinky. Similarly, if you wish to create a longitudinal wave in a slinky, then the first coil of the slinky must be displaced in a direction parallel to the entire slinky.

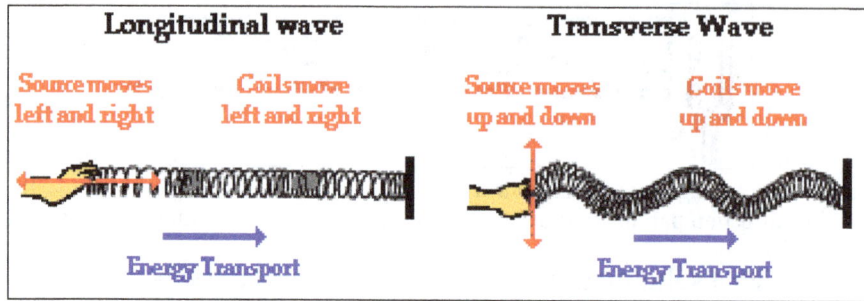

The subsequent direction of motion of individual particles of a medium is
the same as the direction of vibration of the source of the disturabance.

Electromagnetic versus Mechanical Waves

Another way to categorize waves is on the basis of their ability or inability to transmit energy through a vacuum (i.e., empty space). Categorizing waves on this basis leads to two notable categories: electromagnetic waves and mechanical waves.

An electromagnetic wave is a wave that is capable of transmitting its energy through a vacuum (i.e., empty space). Electromagnetic waves are produced by the vibration of charged particles. Electromagnetic waves that are produced on the sun subsequently travel to Earth through the vacuum of outer space. Were it not for the ability of electromagnetic waves to travel to through a vacuum, there would undoubtedly be no life on Earth. All light waves are examples of electromagnetic waves.

A mechanical wave is a wave that is not capable of transmitting its energy through a vacuum. Mechanical waves require a medium in order to transport their energy from one location to another. A sound wave is an example of a mechanical wave. Sound waves are incapable of traveling through a vacuum. Slinky waves, water waves, stadium waves, and jump rope wavesare other examples of mechanical waves; each requires some medium in order to exist. A slinky wave requires the coils of the slinky; a water wave requires water; a stadium wave requires fans in a stadium; and a jump rope wave requires a jump rope.

The above categories represent just a few of the ways in which physicists categorize waves in order to compare and contrast their behaviors and characteristic properties. This listing of categories is not exhaustive; there are other categories as well. The five categories of waves listed here will be used periodically throughout this unit on waves as well as the units on sound and light.

Acoustic Waves

Acoustic waves (also known as sound waves) are a type of longitudinal waves that propagate by means of adiabatic compression and decompression. Longitudinal waves are waves that have the same direction of vibration as their direction of travel. Important quantities for describing acoustic waves are sound pressure, particle velocity, particle displacement and sound intensity. Acoustic waves travel with the speed of sound which depends on the medium they're passing through.

Wave Properties

Acoustic waves are longitudinal waves that exhibit phenomena like diffraction, reflection and interference. Sound waves however don't have any polarization since they oscillate along the same direction as they move.

Phase

In a travelling wave pressure and particle velocity are in phase, which means the phase angle between the two quantities is zero.

This can be easily proven using the ideal gas law,

$$pV = nRT$$

where:

p is pressure in Pa

V is volume in m³

n is amount in mol

R is the universal gas constant with value 8.314472(15) $\dfrac{\text{J}}{\text{mol K}}$

Consider a volume V. As an acoustic wave propagates through the volume, adiabatic compression and decompression occurs. For adiabatic change the following relation between volume V of a parcel of fluid and pressure p holds:

$$\frac{\partial V}{V_m} = \frac{-1}{\gamma} \frac{\partial p}{P_m}$$

where:

γ is the adiabatic index without unit.

As a sound wave propagates through a volume, displacement of particles occurs along the wave propagation direction.

$$\frac{\partial x}{V_m} A = \frac{\partial V}{V_m} = \frac{-1}{\gamma} \frac{\partial p}{P_m}$$

where:

A is cross-sectional area in m².

From this equation it can be seen that when pressure is at its maximum, displacement reaches zero. As mentioned before, the oscillating pressure for a rightward travelling wave can be given by:

$$p = p_0 cos(\omega t - kx)$$

Since displacement is maximum when pressure is zero there is a 90 degrees phase difference, so displacement is given by:

$$x = x_0 sin(\omega t - kx)$$

Particle velocity is the first derivative of particle displacement. Differentiation of a sine gives a cosine again:

$$u = u_0 cos(\omega t - kx)$$

During adiabatic change, temperature changes with pressure as well following:

$$\frac{\partial T}{T_m} = \frac{\gamma - 1}{\gamma} \frac{\partial p}{P_m}$$

This fact is exploited within the field of thermoacoustics.

Propagation Speed

The propagation speed of acoustic waves is given by the speed of sound. In general, the speed of sound c is given by the Newton-Laplace equation:

$$c = \sqrt{\frac{C}{\rho}}$$

where:

C is a coefficient of stiffness, the bulk modulus (or the modulus of bulk elasticity for gas mediums),

ρ is the density in kg/m³.

Thus the speed of sound increases with the stiffness (the resistance of an elastic body to deformation by an applied force) of the material, and decreases with the density. For general equations of state, if classical mechanics is used, the speed of sound c is given by:

$$c^2 = \frac{\partial p}{\partial \rho}$$

where differentiation is taken with respect to adiabatic change.

where p is the pressure and ρ is the density.

Interference

Interference is the addition of two or more waves that results in a new wave pattern. Interference of sound waves can be observed when two loudspeakers transmit the same signal. At certain locations constructive interference occurs, doubling the local sound pressure. And at other locations destructive interference occurs, causing a local sound pressure of zero pascals.

Standing Wave

A standing wave is a special kind of wave that can occur in a resonator. In a resonator superposition of the incident and reflective wave occurs, causing a standing wave. Pressure and particle velocity are 90 degrees out of phase in a standing wave.

Consider a tube with two closed ends acting as a resonator. The resonator has normal modes at frequencies given by:

$$f = \frac{Nc}{2d} \qquad N \in \{1, 2, 3, \ldots\}$$

where:

c is the speed of sound in m/s.

d is the length of the tube in m.

At the ends particle velocity becomes zero since there can be no particle displacement. Pressure however doubles at the ends because of interference of the incident wave with the reflective wave. As pressure is maximum at the ends while velocity is zero, there is a 90 degrees phase difference between them.

Reflection

An acoustic travelling wave can be reflected by a solid surface. If a travelling wave is reflected, the reflected wave can interfere with the incident wave causing a standing wave in the near field. As a consequence, the local pressure in the near field is doubled, and the particle velocity becomes zero.

Attenuation causes the reflected wave to decrease in power as distance from the reflective material increases. As the power of the reflective wave decreases compared to the power of the incident wave, interference also decreases. And as interference decreases, so does the phase difference between sound pressure and particle velocity. At a large enough distance from the reflective material, there is no interference left anymore. At this distance one can speak of the far field.

The amount of reflection is given by the reflection coefficient which is the ratio of the reflected intensity over the incident intensity:

$$R = \frac{I_{reflected}}{I_{incident}}.$$

Absorption

Acoustic waves can be absorbed. The amount of absorption is given by the absorption coefficient which is given by:

$$\alpha = 1 - R^2$$

where:

α is the absorption coefficient without a unit

R is the reflection coefficient without a unit

Often acoustic absorption of materials is given in decibels instead.

Measurement

Sound pressure can be measured directly using a microphone. Particle velocity can be measured directly using a particle velocity probe. It is also possible to measure the quantities indirectly using the opposite instrument. Sound intensity can be measured using different combinations:

- Microphone and particle velocity probe (p-u probe).

- Two microphones (p-p probe).

- Two particle velocity probes (u-u probe).

The sound pressure is measured in pascal, the particle velocity in meters per second, and the sound intensity in watts. Often these quantities are measured as a level in decibels relative to a certain quantity.

The sound pressure level is given by:

$$L_p = 10\log_{10}\left(\frac{p_{rms}^2}{p_{ref}^2}\right)$$

where

p_{rms} is the root-mean square pressure in Pa

p_{ref} is the reference value of $2*10^{-5}$ Pa

The particle velocity level is given by:

$$L_u = 10\log_{10}\left(\frac{u_{rms}^2}{u_{ref}^2}\right)$$

where:

is the root-mean square particle velocity in m/s

is the reference value of $5*10^{-8}$ m/s

The sound intensity level is given by

where:

is the root-mean square sound intensity in W

is the reference value of $1*10^{-12}$ W.

Surface Acoustic Waves

Surface-acoustic waves (SAWs) are sound waves that travel parallel to the surface of an elastic material, with their displacement amplitude decaying into the material so that they are confined to within roughly one wavelength of the surface.

In a piezoelectric material such as gallium arsenide or quartz, the mechanical deformation associated with the SAW produces electric fields. To a good approximation, the electric fields do not affect the propagation of the mechanical wave, so the result is a variation in electrostatic potential that travels along with the SAW. Any thin layers of surface metal or conductive regions in the material affect the electrostatic potential variation around them, but the mechanical SAW propagates largely unaffected.

SAW Transducers

A SAW can be generated by applying a suitable oscillating signal to a suitably designed set of surface gates. A schematic diagram of a typical SAW transducer is shown below that consists of many pairs of interdigitated electrodes forming a grating-like structure, with the pitch of the transducer giving the SAW wavelength. By grounding one side of the transducer and applying a signal at a frequency given by the SAW velocity (roughly 2700 m/s for GaAs) divided by the pitch of the transducer, a SAW can be generated.

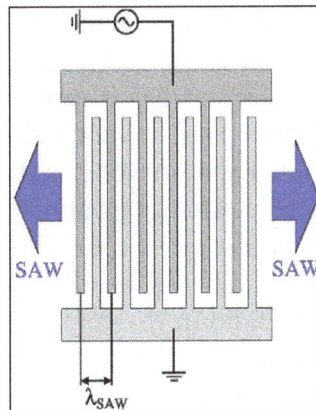

SAW transducer consisting of interdigitated electrodes.

Electron-beam lithography allows sub-micrometre pitch transducers to be fabricated, enabling SAWs with frequencies in excess of 3 GHz to be generated on GaAs. Some pictures of SAW transducers are shown below:

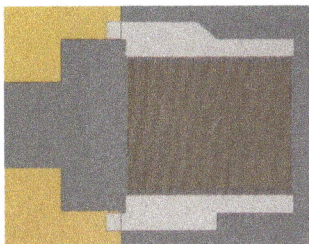

Transducer for generating SAWs with a wavelength of approximately 3 microns.

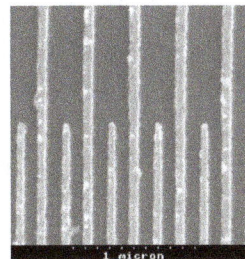

SEM image of part of a transducer for generating SAWs with a wavelength of approximately 0.4 microns.

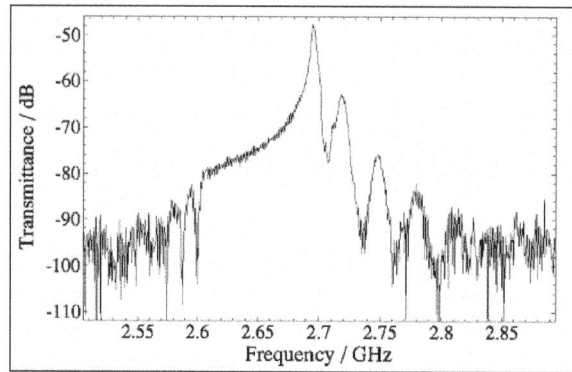

Transmittance between two SAW transducers versus frequency.

The reverse of the process used to generate SAWs can be used to detect them: when a SAW passes underneath a SAW transducer of the appropriate pitch, an alternating potential is generated across the transducer. A graph of the transmittance between two SAW transducers on GaAs, 4 mm apart, is shown below and exhibits a sharp peak corresponding to the generation of a SAW. Such a device can be used as a band-pass filter, and SAW devices are also used for convolvers and delay lines.

Wafers

The high-frequency single-electron-transport devices that have been fabricated to date used semiconductor wafers that contain a two-dimensional electron gas. The wafers consist of layers of GaAs and typically $Al_{0.33}Ga_{0.67}as$ grown by molecular-beam epitaxy. The structure of a typical wafer is shown on the left below, and its calculated conduction band and electron wave-function is shown on the right. We see that the conduction-band offset between the two materials enables potential barriers to be formed, and by doping the wafer appropriately during the growth it is possible to bend the conduction bands so that a narrow potential well is formed that is below the chemical potential and populated with electrons. If the well is sufficiently narrow, and the electron density sufficiently low, then only the first sub-band in the confinement direction will be occupied. The result is an electron gas that is dynamically two-dimensional: the degree of freedom in the confinement direction is not accessible for reasonable electron energies.

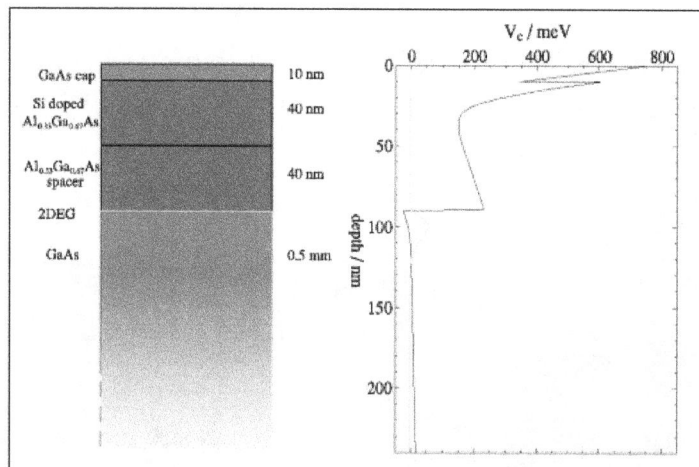

Left: the structure of a typical wafer. Right: its conduction band (black) and electron wave function (blue).

A wide variety of different wafer structures, with different materials, can be grown using molecular-beam epitaxy. Wafers can be grown that have more than one 2DEG in parallel, or with a hole gas, or with an empty quantum well, and n+ GaAs can be used to provide a gate within the wafer.

To make electrical contact to a 2DEG, a mixture of gold, germanium and nickel is deposited on the surface of the wafer and then heated so that it diffuses down into the wafer. This dopes the wafer very heavily so that it conducts, forming an Ohmic contact, and thin gold wires can be attached to the material on the surface.

To pattern a 2DEG, metal electrodes ("gates") can be deposited on the surface. When sufficient negative potential is applied to the gates, the conduction band beneath the gates can be raised so far that the energy of the lowest-lying state in the potential well is brought above the chemical potential and the 2DEG becomes depleted there. Alternatively, the 2DEG can be depleted by etching sufficiently far down into the doped layer of the wafer.

Acoustic Wave Equation

In physics, the acoustic wave equation governs the propagation of acoustic waves through a material medium. The form of the equation is a second order partial differential equation. The equation describes the evolution of acoustic pressure p or particle velocity u as a function of position x and time t. A simplified form of the equation describes acoustic waves in only one spatial dimension, while a more general form describes waves in three dimensions.

For lossy media, more intricate models need to be applied in order to take into account frequency-dependent attenuation and phase speed. Such models include acoustic wave equations that incorporate fractional derivative terms.

In One Dimension

Equation: The wave equation describing sound in one dimension (position x) is;

$$\frac{\partial^2 p}{\partial x^2} - \frac{1}{c^2}\frac{\partial^2 p}{\partial t^2} = 0,$$

where p is the acoustic pressure (the local deviation from the ambient pressure), and where c is the speed of sound.

Solution: Provided that the speed c is a constant, not dependent on frequency (the dispersionless case), then the most general solution is;

$$p = f(ct - x) + g(ct + x)$$

where f and g are any two twice-differentiable functions. This may be pictured as the superposition of two waveforms of arbitrary profile, one (f) travelling up the x-axis and the other (g) down the x-axis at the speed c. The particular case of a sinusoidal wave travelling in one direction is obtained by choosing either f or g to be a sinusoid, and the other to be zero.

$$p = p_0 \sin(\omega t \mp kx).$$

where ω is the angular frequency of the wave and k is its wave number.

Derivation

The wave equation can be developed from the linearized one-dimensional continuity equation, the linearized one-dimensional force equation and the equation of state.

The equation of state (ideal gas law):

$$PV = nRT$$

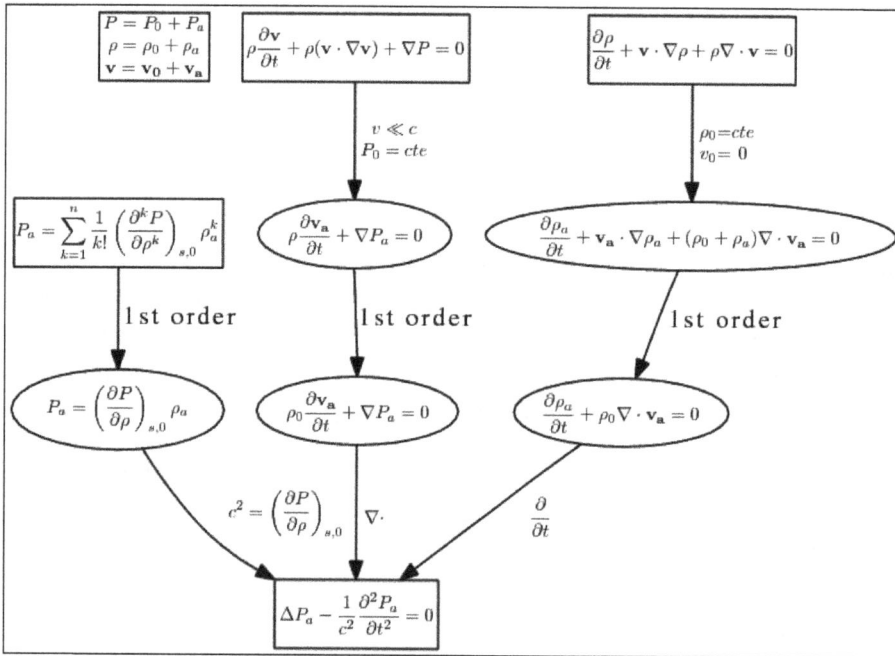

Derivation of the acoustic wave equation

In an adiabatic process, pressure P as a function of density ρ can be linearized to,

$$P = C\rho$$

where C is some constant. Breaking the pressure and density into their mean and total components and noting that $C = \dfrac{\partial P}{\partial \rho}$,

$$P - P_0 = \left(\frac{\partial P}{\partial \rho} \right)(\rho - \rho_0).$$

The adiabatic bulk modulus for a fluid is defined as:

$$B = \rho_0 \left(\frac{\partial P}{\partial \rho} \right)_{adiabatic}$$

which gives the result:

$$P - P_0 = B \frac{\rho - \rho_0}{\rho_0}.$$

Condensation, s, is defined as the change in density for a given ambient fluid density.

$$s = \frac{\rho - \rho_0}{\rho_0}$$

The linearized equation of state becomes:

$$p = Bs,$$

where p is the acoustic pressure ($P - P_0$).

The continuity equation (conservation of mass) in one dimension is:

$$\frac{\partial \rho}{\partial t} + \frac{\partial}{\partial x}(\rho u) = 0.$$

Where u is the flow velocity of the fluid. Again the equation must be linearized and the variables split into mean and variable components.

$$\frac{\partial}{\partial t}(\rho_0 + \rho_0 s) + \frac{\partial}{\partial x}(\rho_0 u + \rho_0 s u) = 0$$

Rearranging and noting that ambient density changes with neither time nor position and that the condensation multiplied by the velocity is a very small number:

$$\frac{\partial s}{\partial t} + \frac{\partial}{\partial x} u = 0$$

Euler's Force equation (conservation of momentum) is the last needed component. In one dimension the equation is:

$$\rho \frac{Du}{Dt} + \frac{\partial P}{\partial x} = 0,$$

where D/Dt represents the convective, substantial or material derivative, which is the derivative at a point moving with medium rather than at a fixed point.

Linearizing the variables:

$$(\rho_0 + \rho_0 s)\left(\frac{\partial}{\partial t} + u \frac{\partial}{\partial x}\right) u + \frac{\partial}{\partial x}(P_0 + p) = 0$$

Rearranging and neglecting small terms, the resultant equation becomes the linearized one-dimensional Euler Equation:

$$\rho_0 \frac{\partial u}{\partial t} + \frac{\partial p}{\partial x} = 0.$$

Taking the time derivative of the continuity equation and the spatial derivative of the force equation results in:

$$\frac{\partial^2 s}{\partial t^2} + \frac{\partial^2 u}{\partial x \partial t} = 0,$$

$$\rho_0 \frac{\partial^2 u}{\partial x \partial t} + \frac{\partial^2 p}{\partial x^2} = 0.$$

Multiplying the first by ρ_0, subtracting the two, and substituting the linearized equation of state,

$$-\frac{\rho_0}{B} \frac{\partial^2 p}{\partial t^2} + \frac{\partial^2 p}{\partial x^2} = 0.$$

The final result is:

$$\frac{\partial^2 p}{\partial x^2} - \frac{1}{c^2} \frac{\partial^2 p}{\partial t^2} = 0,$$

where $c = \sqrt{\dfrac{B}{\rho_0}}$ is the speed of propagation.

In Three Dimensions

Equation: Feynman provides a derivation of the wave equation for sound in three dimensions as:

$$\nabla^2 p - \frac{1}{c^2} \frac{\partial^2 p}{\partial t^2} = 0,$$

where ∇^2 is the Laplace operator, p is the acoustic pressure (the local deviation from the ambient pressure), and where c is the speed of sound.

A similar looking wave equation but for the vector field particle velocity is given by:

$$\nabla^2 u - \frac{1}{c^2} \frac{\partial^2 u}{\partial t^2} = 0.$$

In some situations, it is more convenient to solve the wave equation for an abstract scalar field velocity potential which has the form:

$$\nabla^2 \Phi - \frac{1}{c^2} \frac{\partial^2 \Phi}{\partial t^2} = 0,$$

and then derive the physical quantities particle velocity and acoustic pressure by the equations (or definition, in the case of particle velocity):

$$u = \nabla \Phi,$$

$$p = -\rho \frac{\partial}{\partial t} \Phi.$$

Solution: The following solutions are obtained by separation of variables in different coordinate systems. They are phasor solutions, that is they have an implicit time-dependence factor of $e^{i\omega t}$ where $\omega = 2\pi f$ is the angular frequency. The explicit time dependence is given by:

$$p(r,t,k) = \mathrm{Real}\left[p(r,k)e^{i\omega t} \right]$$

Here $k = \omega / c$ is the wave number.

- Cartesian Coordinates:

$$p(r,k) = Ae^{\pm ikr}.$$

- Cylindrical Coordinates:

$$p(r,k) = AH_0^{(1)}(kr) + BH_0^{(2)}(kr).$$

    where the asymptotic approximations to the Hankel functions, when $kr \to \infty$, are

$$H_0^{(1)}(kr) \simeq \sqrt{\frac{2}{\pi kr}}e^{i(kr - \pi/4)},$$

$$H_0^{(2)}(kr) \simeq \sqrt{\frac{2}{\pi kr}}e^{-i(kr - \pi/4)}.$$

- Spherical Coordinates

$$p(r,k) = \frac{A}{r}e^{\pm ikr}.$$

Depending on the chosen Fourier convention, one of these represents an outward travelling wave and the other a nonphysical inward travelling wave. The inward travelling solution wave is only nonphysical because of the singularity that occurs at r=0; inward travelling waves do exist.

Acoustic Noise

Noise is, to a great extent, a purely subjective personal phenomena. Perhaps the best definition of it is as an unwanted sound. Noise does, however, have two basic characteristics. The first is the physical phenomenon which can be measured and thus used in technical specification. The second is the psycho acoustical characteristic which attempts to judge the effect of noise on human beings. In industries that use small cooling fans, fan noise simply interferes with the ability of the people working nearby to concentrate on their work. The factors of greatest importance to the system designer are the psychological influences on the person rather than the physical influences of sound on the human ear.

Sound is perceived and measured as minute pressure fluctuations above and below the ambient pressure. The pressure variations of interest for their psycho acoustical effect vary as much as 13 orders of magnitude. Because of this large range of hearing capability, it is convenient to express these values in decibels. Sound Pressure Level (SPL) which is environmentally dependent, is defined as:

- $SPL = 20 \log (P/P_{Ref})$

where:

- P = pressure

- P_{Ref} = a reference pressure.

In defining the noise generated by a fan, it is best to define the noise emanating from the source. This is called the Sound Power Level and is independent of the environment. Sound Power Level is defined similarly to sound pressure on a logarithmic scale as:

- $PWL = 10 \log (W/W_{Ref})$

where:

- W = acoustic power of the source

- W_{Ref} = an acoustic reference power.

Sound Power Level cannot be measured directly and must be calculated from sound pressure measurements. Sound Power Level, since it is a measurement of noise unaffected by such factors as the fan's distance from the hearer, is used as the basic measurement for comparing noise levels of fans, as well as noise levels at different operating points of the same fan. In practice, another property of noise, its frequency, is also considered. For fans, two types of noise related to frequency are important: wide band noise, in which acoustic energy is continuously distributed over a frequency spectrum; and pure tones, in which the acoustic energy is concentrated over narrow bands in the frequency spectrum.

Since fan noise is predominantly wide band in nature with some pure tones, it is convenient to divide the audible frequency range into bands and to plot the average Sound Power Level in each band. For specification and rating purposes, it is generally acceptable to divide the audible frequency spectrum into eight octave bands, each with an upper limit twice that of the lower limit. These bands are usually designated by their center frequency. Fan noise data is usually plotted as Sound Power level against the octave frequency bands.

Noise Rating Systems

Comair Rotron uses four rating methods for describing the noise levels in the fans it manufactures:

- PSIL: The first system used is Preferred Speech Interference Level. The PSIL is determined as the arithmetic average of the sound pressure level in the three octave bans with center frequencies of 500, 1000 and 2000 Hz. This rating is a good guide to the effect of noise on spoken communications.

- dBA: A second rating system is the "A" weighted sound pressure level (dBA) often used by government agencies in determining compliance with such regulations as the

Occupational Safety and Health Act (OSHA). The dBA rating is determined directly by a sound level meter equipped with a filtering system which de-emphasizes both the low and high frequency portions of the audible spectrum. This measurement is recorded at a distance of 3 feet from the source.

- NPEL: A third rating system is the "A" weighted sound power level reference to a 1 picowatt and expressed in Bels. This is also referred to as the Noise Power Emission Level (NEPL). NEPL was adopted by the Institute of Noise Control Engineering (INCE) as the preferred unit of measure. The INCE "Recommended Practice for Measurement of Noise Emitted by Air Moving Devices (AMDs) for Computer and Business Equipment" is a guideline for the description and control of noise emitted by components. ANSI S12.11 now includes the procedures called for in the INCE Practice. This is the latest and most technically thorough acoustic test procedure available. Comair Rotron does all acoustical testing per INCE and ANSI S12.11-1987.

- Freely Suspended: The fourth rating system used is a method known as Freely Suspended. In this method a fan is suspended from springs in the middle of a Calibrated Reverberate Room. The fan is run at nominal voltage, free delivery, and at a distance of 1 meter. The sound pressure level (dBA) is recorded. (For comparison dBA @ 1 meter + .7778 = dBA @ 3 feet).

Causes of Fan Noise

Since noise in most measuring systems is specified in decibels (dB), it is useful to see how dB changes relate to perceived loudness:

dB Change Apparent Change in Loudness:

- 3dB Just noticeable.

- 5dB Clearly noticeable.

- 10dB Twice (or half) as loud.

Noise emanating from axial fans is a function of many variables and causes:

- Vortex Shedding: This is a broad band noise source generated by air separation from the blade surface and trailing edge. It can be controlled somewhat by good blade profile design, proper pitch angle and notched or serrated trailing blade edges.

- Turbulence: Turbulence is created in the airflow stream itself. It contributes to broad band noise. Inlet and Outlet disturbances, sharp edges and bends will cause increased turbulence and noise.

- Speed: The effect of speed on noise can best be seen through one of the fan laws:

  ○ $dB_1 = dB_2 + 50 \log_{10} (RPM_1 / RPM_2)$

Speed is a major contributor to fan noise. For instance, if the speed of a fan is reduced by 20%, the dB level will be reduced by 5 dB.

- Fan Load: Noise varies as the system load varies. This variation is unpredictable and fan dependent. However, fans are generally quieter when operated near their peak efficiency.

- Structure Vibration: This can be caused by the components and mechanism within the fan, such as residual unbalance, bearings, rotor to stator eccentricity and motor mounting. Motor mounting noise is difficult to define. It should be remembered that cooling fans are basically motors and should be treated as such when mounted.

System Effects on Fan Noise

System disturbances are the biggest cause of fan noise. When a fan is designed for low noise operation, it can be very sensitive to inlet and outlet disturbances caused by card guides, brackets, capacitors, transformers, cables, finger guards, filter assemblies, walls or panels, etc.

When placing a fan in an electronic package, great care should be taken in locating components. Trial and error will be needed to determine the system's effect on noise. Different fan types will react differently in the same system. Common sense and intuition play a large role in the fan/system design.

For instance, if it is necessary to place card guides against the face of the fan for card cooling, the fan may develop a large pure tone if it is done on the inlet side; on the discharge side, the effect may be much less.

Guidelines for Low Noise

System Impedance

This should be reduced to the lowest possible level so that the least noise for the most airflow is obtained. The inlet and outlet ports of a cabinet can make up to between 60 and 80% of the total system impedance, which is much too high for a low-noise result. And, if a large part of the fan's flow potential is used up by the impedance of the inlet and outlet, a larger, faster and noisier fan will be required to provide the necessary cooling.

Flow Disturbance

Obstructions to the airflow must be avoided whenever possible, especially in the critical inlet and outlet areas. When turbulent air enters the fan, noise is generated, usually in discrete tone form, that can be as mush as 10 dB higher and thus cause considerable annoyance.

Fan Speed and Size

Most Comair Rotron fans have several low speed versions. These should be tried and used whenever possible. Various fan sizes should also be explored; quite often a larger, slower fan will be quieter than a smaller, faster fan delivering the same airflow.

Temperature Rise

Airflow is inversely proportional to allowable temperature rise with the system. Therefore, the rT limit placed on a piece of equipment will dictate to a large extent the required flow, and therefore, noise. If the temperature limit can be relaxed even a small amount, a noise reduction may result.

Vibration Isolation

In certain instances, the fan must be isolated from the cabinet to avoid vibration transmission. Because fans operate at a low frequency, and are light in weight, vibration isolators must be soft and flexible. Since the transmission is dependent on the system, trial and error is the best approach to a quiet system/fan interaction. In systems that require 20 CFM or less, noise radiated by the cabinet is the predominant noise. Isolation of the fan is the only practical solution to this type of system noise problem.

Speech Acoustics

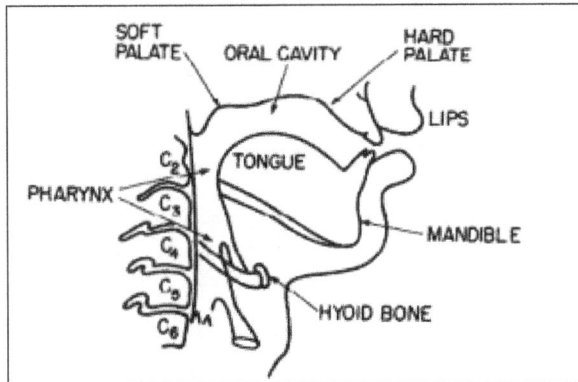

Anatomical diagrams of human vocal tract

The study of speech acoustics has been a growing and evolving field of research for many years. Imaging the vocal tract to study speech production has progressed from x-ray videos of a human subject to MRI scans and computer simulations. These advancements have not only greatened our precision and accuracy in our analysis of speech production but have also made the data retrieval process significantly safer and more comfortable for human subjects. However, our knowledge of speech production is still hindered by the fact that the vocal tract is, for the most part, concealed from the naked eye. With today's modern technology, doctors can examine a wide variety of internal body parts, organs, and cavities via very small cameras attached to tubes or probes. Unfortunately, this advancement fails to benefit speech production research because no accurate data can be gathered from a subject with a long tube inserted into their mouth and down their throat. Therefore, modeling the human vocal tract is both essential to our understanding of how we speak and very difficult.

Speech is produced by forcing air from our lungs through our trachea and the rest of the vocal tract. For some speech sounds, such as vowels, the air pressure causes the vocal folds to vibrate, thus providing the sound waves that we define as speech.

It is the shape of the vocal tract between the glottis (vocal cords) and the lips that determines which speech sounds are produced. The vocal tract constricts and expands in crucial places to change the resonant frequencies associated with a speech sound. All phonemes or distinct units of speech defined for a certain language, have identifying resonant frequencies known as formant frequencies. The movement of one's articulators (i.e lips, tongue, teeth, throat) can change these and result in different sounds.

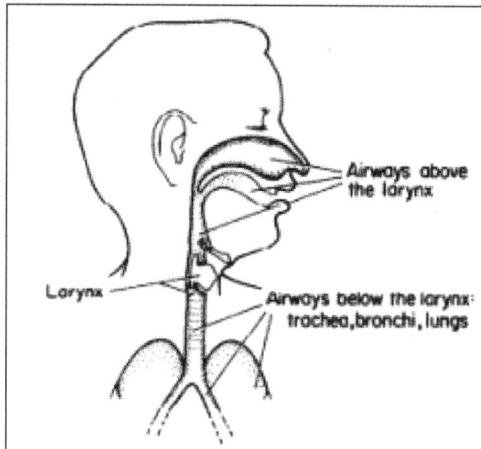

Pitch

Pitch, in music is the position of a single sound in the complete range of sound. Sounds are higher or lower in pitch according to the frequency of vibration of the sound waves producing them. A high frequency (e.g., 880 hertz [Hz; cycles per second]) is perceived as a high pitch and a low frequency (e.g., 55 Hz) as a low pitch.

In Western music, standard pitches have long been used to facilitate tuning among various performing groups. Usually a′ above middle C (c′) is taken as a reference pitch. The current standard pitch of a′ = 440 Hz was adopted in 1939. For some eighty years previous, a′ had been set at 435 Hz. A confusing variety of pitches prevailed until the 19th century, when the continual rise in pitch made some international agreement a matter of practical necessity.

In the mid-17th century the Hotteterres, Parisian instrument makers, remodeled the entire woodwind family, using the Paris organ pitch of about a′ = 415, or a semitone below a′ = 440. This new, or Baroque, pitch, called *Kammerton* ("chamber pitch") in Germany, was one tonebelow the old Renaissance woodwind pitch, or *Chorton* ("choir pitch").

After about 1760 the conventional pitch rose, reaching a′ = 440 by about 1820. By the latter half of the 19th century, it had reached the "Old Philharmonic Pitch" of about a′ = 453. The inconvenience of this high pitch became apparent, for it strained singers' voices and made wind instruments quickly out of date. An international commission met in Paris in 1858–59 and adopted a compromise pitch called diapason normal (known in the United States as "French pitch" or "international pitch") at a′ = 435. England, in 1896, adopted the "New Philharmonic Pitch" at a′ = 439 and, in 1939, adopted the U.S. standard pitch of a′ = 440. In the mid-20th century, pitch again tended to creep upward as some European woodwind builders used the pitch a′ = 444.

When frequency numbers are not used for a particular pitch, say D or B, a system of lowercase and capital letters indicates the octave in which it occurs. The notes in the octave below middle C are indicated by lowercase letters from c to b, the notes of the second octave below middle C are shown as C, D,...B, and the notes of the next lower octave as C′, D′,...B′. Middle C is shown as c′ and the notes in the octave above middle C as d′, e′,...b′. The C above middle C is shown as c″ and the next higher C as c‴.

Absolute, or perfect, pitch is the ability to identify by ear any note at some standard pitch or to sing a specified note, say G#, at will. Fully developed absolute pitch is rare. It appears early in childhood and is apparently an acute form of memory of sounds of a particular instrument, such as the home piano. Some musicians slowly acquire a degree of absolute pitch, if only for the familiar a′ = 440. In general, the ability of humans to process sounds associated with music is due to the development of brain areas that are specialized to be sensitive to pitch; other animals appear to lack this specialization in brain development.

Loudness

Loudness, in acoustics is the attribute of sound that determines the intensity of auditory sensation produced. The loudness of sound as perceived by human ears is roughly proportional to the logarithm of sound intensity: when the intensity is very small, the sound is not audible; when it is too great, it becomes painful and dangerous to the ear. The sound intensity that the ear can tolerate is approximately 10^{12} times greater than the amount that is just perceptible. This range varies from person to person and with the frequency of the sound.

A unit of loudness, called the phon, has been established. The number of phons of any given sound is equal to the number of decibels of a pure 1,000-hertz tone judged by the listener to be equally loud. The decibel scale is objective in that the intensity is defined physically and any intensity can be compared directly with the physically defined reference point. The phon scale is partially subjective in that the judgment of a listener is involved in comparing any arbitrary sound with the physically defined reference in order to establish its loudness in phons. The average result from a large number of people then establishes the definition of equal loudness curves (i.e., curves that show the varying absolute intensities of a pure tone that has the same loudness to the ear at various frequencies).

A third, more-subjective loudness scale involves listener judgment as to what constitutes "doubling" of the loudness of a sound. A tone having a loudness of 40 phons is defined as having a subjective loudness of one sone; a tone judged by the listener to be "twice as loud" would have a loudness of two sones, three times as loud would be three sones, and so forth. As in the case of the definition of the phon, the average values from observations by a large number of people would then define the details of the scale for purposes of classifying and measuring sound levels.

Subjective scales were developed because they tend to be more useful than a totally objective scale in describing how the ear works. In general, the physical sciences and engineering use more-objective scales such as the decibel, while measurements in biological and medical fields tend to use the more-subjective scales.

Difference between Noise and Sound

It is not that hard to discern noise from sound. Sound is a result of the vibrating air in the surroundings. Vibration passes through the air. It creates different levels of air pressure (higher and lower) through air compression and decompression. These variations travel across the air in the form of sound waves which are responsible for the creation of sound. Although these can't be seen, sounds are perceived by the sense of hearing. One of the most pleasant forms of sound is that which is generated by non-hardcore musical instruments.

Noise, on the contrary, is a kind of sound – a remarkably loud one, that is. In this regard, shouts are the perfect examples of noise. It is characterized by its unpleasantness and annoying nature that can even lead to some physical ill effects. Aside from hearing loss, noise can also induce severe cardiovascular symptoms of increased heart rate and can bring about psychological effects that manifest as anxiety, lack of concentration, and profound nervousness.

Moreover, noise is therefore deafening all the time compared to sound which can easily be handled or listened to by everyone. This is the reason why noise is a type of sound that is less or least desired. When there's constant, loud chattering inside the classroom, or when there's a powerful rock concert held in your neighborhood, one might easily get freaked out because of the terrible noises these situations create.

And so, many means have been employed to reduce or control noise through engineering modification of the workplace or environment. Earplugs are also used in noisy places to avoid hearing problems. Nevertheless, staying away from sources of loud noise is the single best means of prevention.

In terms of vibration characteristics, sound is more regular while noise has a more irregular quality that is constantly fluctuating in a seemingly uncontrolled manner. Decibel (dB), the measure of sound strength or loudness, is used to gauge sound or noise intensity. Obviously, the sound with a higher dB value is louder. Stronger sounds that start from 120 dB can already be considered as noise. Take note, a crying baby can already reach as high as 115 dB. Most safe sounds are less than 100 dB. Hertz (Hz) is another unit that measures sound frequency. The higher the frequency, the higher the pitch, and the more annoying the sound becomes.

Wave Propagation

Wave is a disturbance that transfers energy through medium or space with negligible or no amount of mass transfer. There are various types of waves which render many different types of services. Electromagnetic waves are widely used in engineering applications. We use waveforms in various types of applications such as wireless communication, Radar, Space Exploration, Marine, Radio navigation, Remote sensing etc. Among these applications, some uses guided medium for sending waves whereas some make use of the unguided medium.

Electromagnetic Waves are generated by the radiated power from the current carrying conductor. In conductors, a part of the generated power escapes and propagates into free space in the form of Electromagnetic wave, which has a time-varying electrical field, magnetic field, and direction of propagation orthogonal to each other.

Radiated from an isotropic transmitter, these wave travels through different paths to reach the receiver. The path taken by the wave to travel from the transmitter and reach the receiver is known as Wave Propagation.

Electromagnetic (EM) or Radio Wave Propagation

When the isotropic radiator is used for transmission of EM waves we get spherical wavefronts as shown in the figure because it radiates EM waves uniformly and equally in all directions. Here the center of the sphere is the radiator while the radius of the sphere is R. Clearly, all the points at the distance R, lying on the surface of the sphere have equal power densities.

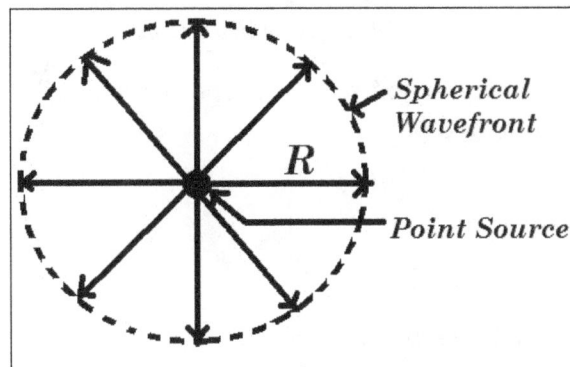

Spherical Wavefront.

The E waves travel in the free space with the velocity of light .i.e. c = But EM waves travel through another medium the speed reduces. The speed of EM waves in any medium other than free space is given by,

$$v = \frac{c}{\sqrt{\varepsilon_r}},$$

where c is the velocity of light and is relative permittivity of the medium.

EM waves transmit energy by absorption and re-emission of wave energy by the atoms in the medium. The atoms absorb the wave energy, undergo vibrations and pass the energy by re-emission of EM of the same frequency. The optical density of the medium affects the propagation of EM waves.

Wave Propagation Equation

Waves take many routes on their way to reach the receiver. Many parameters decide the path taken by the wave such as heights of the transmitting and receiving antennas, the angle of launching at transmitting end, the frequency of operation polarization etc.

Many of the properties of the waves get modified during propagation such as reflection, refraction, diffraction etc, due to the variation of parameters of propagating media like conductivity, permittivity, permeability, and characteristics of obstructing objects.

Generally, when power is radiated in the free space, the wave energy may be radiated or absorbed by the objects in the medium. So while transmitting a wave through a medium it is essential to calculate the loss that can be caused to the wave. This loss is called Radio transmission loss, which is based on inverse square law of optics and is calculated as the ratio of the radiated power to the received power.

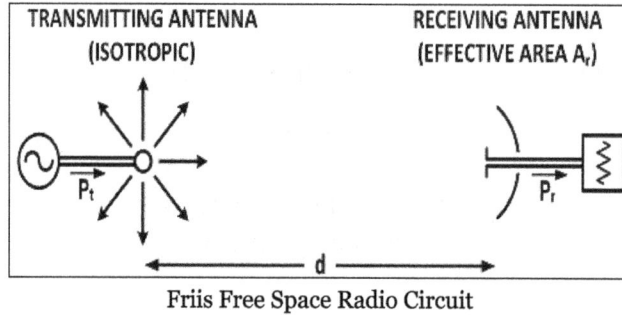

Friis Free Space Radio Circuit

As we know that when an isotropic transmitter is used, power is distributed equally, average power can be expressed in terms of radiated power as,

$$P_{avg} = \frac{\frac{P_{rad}}{4\pi r^2} W}{m^2}$$

The directivity of a test antenna is given by,

$$\left(G_{D\max}\right)_t = \frac{P_{d\max}}{\frac{P_{rad}}{4\pi r^2}}$$

$$\therefore P_{d\max} = G_{D\max} \cdot \frac{P_{rad}}{4\pi r^2} \rightarrow (2)$$

Assume that the receiving antenna receives all the generated power from the radio waves without any loss. Let be the maximum power received by the receiver antenna under a matched load condition. When is the effective aperture of the receiving antenna, we can write as,

$$P_{rec} = P_{d\max}\left(A_e\right)_r$$

$$\therefore P_{rec} = \left(G_{D\max}\right)_t \cdot \frac{P_{rad}}{4\pi r^2}\left(A_e\right)_r$$

In general, the directivity and effective aperture area for any antenna is related as,

$$G_{D\max} = \frac{4\pi}{\lambda^2}\left(A_e\right)$$

Let be the directivity of the receiving antenna. Then,

Let $\left(G_{D\max}\right)_r$ be the directivity of the receiving antenna. Then,

$$\left(G_{D\max}\right)_r = \frac{4\pi}{\lambda^2}\left(A_e\right)_r$$

$$\therefore \left(A_e\right)_r = \frac{\lambda^2}{4\pi}\left(G_{D\max}\right)_r$$

Substituting the value in $P_{rec} = \left(G_{D\max}\right)_t \cdot \dfrac{P_{rad}}{4\pi r^2}\left(A_e\right)_r$ we get,

$$P_{rec} = \left(G_{D\max}\right)_t \cdot \dfrac{P_{rad}}{4\pi r^2}\left[\dfrac{\lambda^2}{4\pi}\left(G_{D\max}\right)_r\right]$$

$$\therefore \dfrac{P_{rec}}{P_{rad}} = \left(G_{D\max}\right)_t \left(G_{D\max}\right)_r \left(\dfrac{\lambda}{4\pi r}\right)^2$$

This equation is known as the Fundamental Equation for free space Propagation, also known as Friss free space equation. The factor $(\lambda/4\pi r)^2$ is called free space path loss which indicates the loss of the signal. Path loss can be expressed as,

$$P_{Loss} = 10\log_{10}\left(\dfrac{4\pi r}{\lambda}\right)^2 \text{ dB}$$

We can express the equation $\dfrac{P_{rec}}{P_{rad}} = \left(G_{D\max}\right)_t \left(G_{D\max}\right)_r \left(\dfrac{\lambda}{4\pi r}\right)^2$ in dB as,

$$10\log_{10}\left(\dfrac{P_{rec}}{P_{rad}}\right) = 10\log_{10}\left(G_{D\max}\right)_t + 10\log_{10}\left(G_{D\max}\right)_r + 10\log_{10}\left[\left(\dfrac{\lambda}{4\pi r}\right)^2\right]$$

Received power can be expressed as,

$$P_{rec(dB)} = P_{rad(dB)} + \left(G_{D\max}\right)_{r(dB)} + \left(G_{D\max}\right)_{t(dB)} - L_{s(dB)}$$

$$\text{Where } L_{s(dB)} = 10\log_{10}\left[\left(\dfrac{\lambda}{4\pi r}\right)^2\right] = 20\log_{10}\left(\dfrac{\lambda}{4\pi r}\right)$$

Which, on simplification is given as,

$$L_{s(dB)} = 32.45 + 20\log_{10} r + 20\log_{10} f,$$

Here distance r is expressed in kilometer while frequency f is expressed in MHz. This indicates loss due to wave spreading taking place when it propagates out of the source.

Types of Wave Propagation

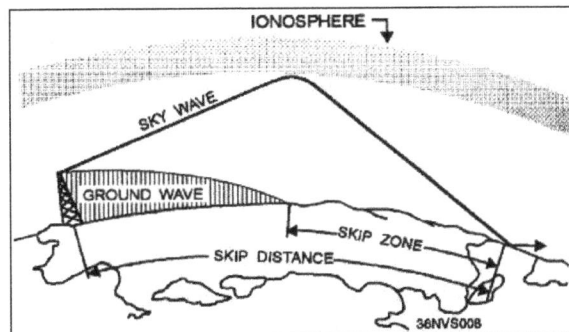

Wave Propagation

The electromagnetic waves or radio waves propagation, passing through the environment of the earth depend not only on the properties of themselves but also on the properties of the environment. There are different paths of propagation by which the transmitted waves can reach the receiver. All these modes depend on the frequency of operation, the distance between transmitter and receiver etc.

- The waves that propagate near the earth's surface are called ground waves. This type of propagation is possible when the transmitting and receiving antenna both are closed to the earth's surface.

- The ground waves which travel without any reflection are called Direct waves or Space waves.

- The ground waves which propagate to the receiving antenna through the reflection from the earth's surface are called Ground Reflected waves or Surface waves.

- The waves which reach the receiving antenna due to scattering and reflection by the ionization in the upper atmosphere are called Skywaves.

- The waves which are reflected or scattered in the troposphere before reaching antenna are called troposphere waves.

Ground Wave or Surface Wave Propagation

A ground wave travels along the surface of the earth. These waves are vertically polarized. So, vertical antennas are useful for these waves. If a horizontally polarized wave is propagated as a ground wave, due to the conductivity of the earth, the electric field of the wave gets short-circuited.

As the ground wave travels away from the transmitting antenna it gets attenuated. To minimize this loss the transmission path must be over the ground with high conductivity. With respect to this condition, sea water should be the best conductor but it was observed that large storage of water in ponds, sandy or rocky soil shows maximum losses.

Hence, high power low-frequency transmitters, using ground wave propagations, are preferably located on ocean fronts. As ground losses increase rapidly with frequency, this propagation is used practically for signals up to frequency 2 MHz only.

For medium wave broadcast although ground waves are preferred some energy is transmitted to the ionosphere. But during day time the energy is completely absorbed by the ionosphere and during night time ionosphere reflects energy back to earth. So all the broadcast signal received during day time are due to ground wave only.

The maximum range of ground wave propagation not only depends on the frequency but also on the power of the transmitter. As ground waves pass over the surface of the earth they are also called the Surface wave.

Sky Wave Propagation

Every long radio communication of medium and high frequencies are conducted using skywave propagation. In this mode reflection of EM waves from the ionized region in the upper part of the atmosphere of the earth is used for transmission of waves to longer distances.

This part of the atmosphere is called ionosphere which is at about 70-400 km height. Ionosphere reflects back the EM waves if the frequency is between 2 to 30 MHz's. Hence, this mode of propagation is also called as Short wave propagation.

Using sky wave propagation point to point communication over long distances is possible. With the multiple reflections of sky waves, global communication over extremely long distances is possible.

But a drawback is that the signal received at the receiver has faded due to a large number of waves following a large number of different paths to reach the receiving point.

Space Wave Propagation

When we are dealing with EM waves of frequency between 30 MHz to 300 MHz, then space wave propagation is useful. Here properties of Troposphere are used for transmission.

When operating in space wave propagation mode, the wave reaches the receiving antenna directly from the transmitter or after reflection from troposphere which is present at about 16km above the earth surface. Hence space wave mode consists of two components .i.e. direct wave and indirect wave.

Though these components are transmitted at the same time with the same phase they may reach within the phase or out of phase with each other at the receiver end depending on the different path lengths. Thus, at the receiver side signal strength is the vector sum of strengths of direct and indirect waves.

The space wave propagation mode is used for propagation of very high frequencies.

Which of the Propagation is used for Short Wave Broadcasting

Short wave broadcasting usually takes place in the frequency range of 1.7 – 30 MHz. the frequencies in this range are propagated through Skywave propagation mode.

Depending on the frequency or wavelength the electromagnetic waves produce different affected in various materials and devices. Hence, the different parts of the electromagnetic spectrum are utilized for various applications.

Acoustic Tweezers

Acoustic tweezers are used to manipulate the position and movement of very small objects with sound waves. The technology works by controlling the position of acoustic pressure nodes that draw objects to specific locations of a standing acoustic field. The target object must be considerably smaller than the wavelength of sound used, and the technology is typically used to manipulate microscopic particles.

Acoustic waves have been proven safe for biological objects, making them ideal for biomedical applications. Recently, applications for acoustic tweezers have been found in manipulating sub-millimetre objects, such as flow cytometry, cell separation, cell trapping, single-cell manipulation, and nanomaterial manipulation. The use of one-dimensional standing waves to manipulate small particles was first reported in the 1982 research article "Ultrasonic Inspection of Fiber Suspensions".

Method

In a standing acoustic field, objects experience an acoustic-radiation force that moves them to specific regions of the field. Depending on an object's properties, such as density and compressibility, it can be induced to move to either acoustic pressure nodes (minimum pressure regions) or pressure antinodes (maximum pressure regions). As a result, by controlling the position of these nodes, the precise movement of objects using sound waves is feasible. Acoustic tweezers do not require expensive equipment or complex experimental setups.

Fundamental Theory

Particles in an acoustic field can be moved by forces originating from the interaction among the acoustic waves, fluid, and particles. These forces (including acoustic radiation force, secondary field force between particles, and Stokes drag force) create the phenomena of acoustophoresis, which is the foundation of the acoustic tweezers technology.

Acoustic Radiation Force

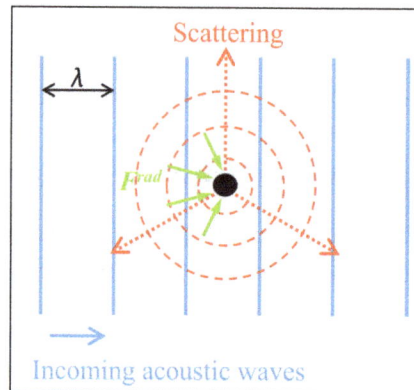

Acoustic radiation force on a small particle

When a particle is suspended in the field of a sound wave, an acoustic radiation force that has risen from the scattering of the acoustic waves is exerted on the particle. This was first modeled and analyzed for incompressible particles in an ideal fluid by Louis King in 1934. Yosioka and Kawasima calculated the acoustic radiation force on compressible particles in a plane wave field in 1955. Gorkov summarized the previous work and proposed equations to determine the average force acting on a particle in an arbitrary acoustical field when its size is much smaller than the wavelength of the sound. Recently, Bruus revisited the problem and gave detailed derivation for the acoustic radiation force.

As shown in figure, the acoustic radiation force on a small particle results from a non-uniform flux of momentum in the near-field region around the particle, $F^{rad} = -\nabla U$, which is caused by the incoming acoustic waves and the scattering on the surface of the particle when acoustic waves propagate through it. For a compressible spherical particle with a diameter much smaller than the wavelength of acoustic waves in an ideal fluid, the acoustic radiation force can be calculated by $F^{rad} = -\nabla U$, where U is a given quantity, also called acoustic potential energy. The acoustic potential energy is expressed as:

$$U = V_0 \left(\frac{\overline{p_{in}^2}}{2\rho_f c_f^2} f_1 - \frac{3\rho_f \overline{v_{in}^2}}{4} f_2 \right)$$

where

- V_0 is the particle volume,

- p_{in} is the acoustic pressure,

- v_{in} is the velocity of acoustic particles,

- ρ_f is the fluid mass density,

- c_f is the speed of sound of the fluid,

- $\overline{<*>}$ is the time-average term, $\frac{1}{T} \int_0^T (*) dt$

The coefficients f_1 and f_2 can be calculated by $f_1 = 1 - \frac{\rho_f c_f^2}{\rho_p c_p^2}$ and $f_2 = \frac{2(\rho_p - \rho_f)}{2\rho_p + \rho_f}$

where

- ρ_p is the mass density of the particle,

- c_p is the speed of sound of the particle.

Acoustic Radiation Force in Standing Waves

The standing waves can form a stable acoustic potential energy field, so they are able to create stable acoustic radiation force distribution, which is desirable for many acoustic tweezers applications. For one-dimension planar standing waves, the acoustic fields are given by:

$$A_{in}(x,t) = \frac{-P_0}{\rho_f \omega} \sin(kx) \sin(\omega t),$$

$$p_{in}(x,t) = P_0 \cos(kx) \sin(\omega t),$$

$$\mathbf{v_{in}}(x,t) = \frac{-P_0}{\rho_f c_f} \sin(kx) \cos(\omega t)) \mathbf{e_x},$$

where

- A_{in} is the displacement of acoustic particle,

- P_0 is the acoustic pressure amplitude,

- ω is the angular velocity,

- k is the wave number.

With these fields, the time-average terms can be obtained. These are:

$$\overline{p_{in}^2} = \frac{1}{2}P_0^2\cos^2(kx),$$

$$\overline{v_{in}^2} = \frac{P_0^2}{2\rho_f^2 c_f^2}\sin^2(kx),$$

Thus, the acoustic potential energy is:

$$U = V_0\frac{P_0^2}{4\rho_f c_f^2}[\cos^2(kx)f_1 - \frac{3}{2}\sin^2(kx)f_2],$$

Then, the acoustic radiation force is found by differentiation:

$$F_x^{rad} = -\partial_x U = V_0 k E_{ac}\sin(2kx)\Phi$$

$$E_{ac} = \frac{P_0^2}{4\rho_f c_f^2}, \Phi = f_1 + \frac{3}{2}f_2 = \frac{5\rho_p - 2\rho_f}{2\rho_p + \rho_f} - \frac{\rho_f c_f^2}{\rho_p c_p^2}$$

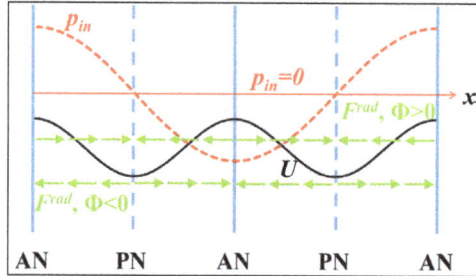

Positions of pressure nodes (PN) and antinodes (AN) along acoustic pressure waveform.

where:

- E_{ac} is the acoustic energy density, and

- Φ is acoustophoretic contrast factor.

The term $\sin(2kx)$ shows that the radiation force period is one-half of the pressure period. Also, the contrast factor can be positive or negative depending on the properties of particles and fluid. For positive value of Φ, the radiation force points from the pressure antinodes to the pressure nodes, as shown in Figure, and the particles will be pushed to the pressure nodes.

Secondary Acoustic Forces

When multiple particles in a suspension are exposed to a standing wave field, they will not only experience acoustic radiation force, but also secondary acoustic forces caused by waves scattered by other particles. The inter-particle forces are sometimes called Bjerknes forces. A simplified equation for the inter-particle forces of identical particles is:

$$F_B(x) = 4\pi a^6[\frac{(\rho_p - \rho_f)^2(3\cos^2\theta - 1)}{6\rho_f d^4}v_{in}^2(x) - \frac{\omega^2\rho_f}{9d^2}(\frac{1}{\rho_p c_p^2} - \frac{1}{\rho_f c_f^2})^2 p_{in}^2(x)]$$

where:

- a is the radius of the particle,
- d is the distance between the particles,
- θ is the angle between the central line of the particles and the direction of propagation of the incident acoustic wave.

The sign of the force represents its direction: a negative sign for an attractive force, and a positive sign for a repulsive force. The left side of the equation depends on the acoustic particle velocity amplitude $v_{in}(x)$ and the right side depends on the acoustic pressure amplitude $p_{in}(x)$. The velocity-dependent term is repulsive when particles are aligned with wave propagation ($\Theta=0°$), and negative when perpendicular to wave propagation ($\Theta=90°$). The pressure-dependent term is unaffected by the particle orientation and is always attractive. In the case of a positive contrast factor, the velocity-dependent term diminishes as particles are driven to the velocity node (pressure antinode), as in the case of air bubbles and lipid vesicles. In a similar way, the pressure-dependent term diminishes as particles are driven towards the pressure node (velocity antinode), as are most solid particles in aqueous solutions.

The influence of the secondary forces is usually very weak, and only has an affect when the distance between particles is very small. It becomes important in aggregation and sedimentation applications, where particles are initially gathered in nodes by the acoustic radiation force. As interparticle distances become smaller the secondary forces assist in a further aggregation until the clusters become heavy enough for sedimentation to begin.

Acoustic Streaming

Acoustic streaming is a steady flow generated by a nonlinear effect in an acoustic field. Depending on the mechanisms, the acoustic streaming can be categorized into two general types, Eckert streaming and Rayleigh streaming. Eckert streaming is driven by a time-average momentum flux created when high-amplitude acoustic waves propagate and attenuate in fluid. Rayleigh streaming, also called "boundary driven streaming", is forced by a shear viscosity close to a solid boundary. Both of the driven mechanisms come from a time-average nonlinear effect.

A perturbation approach is used to analyze the phenomenon of nonlinear acoustic streaming. The governing equations for this problem are mass conservation and Navier-Stokes equations:

$$\partial_t \rho = -\nabla \cdot (\rho v),$$

$$\rho[\partial_t v + (v \cdot \nabla)v] = -\nabla p + \mu \nabla^2 v + \beta \mu \nabla (\nabla \cdot v)),$$

where:

- ρ is the density of fluid,
- v is the velocity of fluid particle,
- p is the pressure,
- μ is the dynamic viscosity of fluid,
- β is the viscosity ratio.

The perturbation series can be written as $p = p_0 + p_1 + p_2$, $v = 0 + v_1 + v_2$, $\rho = \rho_0 + \rho_1 + \rho_2$, which are diminishing series with the higher-order terms much smaller than the lower-order ones.

The liquid is quiescent and homogeneous at its zero-order state. Substituting the perturbation series into the mass conservation and Navier-Stokes equation and using the relation of $p_1 = c_f^2 \rho_1$, the first-order equations can be obtained by collecting terms in first-order,

$$\partial_t p_1 = -\rho_0 c_f^2 \nabla \cdot v_1,$$

$$\rho_0 \partial_t v_1 = -\nabla p_1 + \mu \nabla^2 v_1 + \beta \mu \nabla (\nabla \cdot v_1).$$

Similarly, the second-order equations can be found as well,

$$\partial_t \rho_2 = -\rho_0 \nabla \cdot v_2 - \nabla \cdot (\rho_1 v_1),$$

$$\rho_0 \partial_t v_2 = -\nabla p_2 + \mu \nabla^2 v_2 + \beta \mu \nabla (\nabla \cdot v_2) - \rho_1 \partial_t v_1 - \rho_0 (v_1 \cdot \nabla) v_1.$$

For the first-order equations, taking the time derivation of the Navier-Stokes equation and inserting the mass conservation, a combined equation can be found:

$$\frac{1}{c_f^2} \partial_t^2 p_1 = [1 - \frac{(1-\beta)\mu}{\rho_0 c_f^2} \partial_t] \nabla^2 p_1.$$

This is an acoustic wave equation with viscous attenuation. Physically, and can be interpreted as the acoustic pressure and the velocity of the acoustic particle.

The second-order equations can be considered as governing equations used to describe the motion of fluid with mass source $[-\nabla \cdot (\rho_1 v_1))]$ and force source $[-\rho_1 \partial_t v_1 - \rho_0 (v_1 \cdot \nabla) v_1]$. Generally, the acoustic streaming is a steady mean flow, where the response time scale is much smaller than the one of the acoustic vibration. The time-average term v_2 is normally used to present the acoustic streaming. By using $\overline{\partial_t \rho_2} = 0,$, the time-average second-order equations can be obtained:

$$\rho_0 \nabla \cdot \overline{v_2} = -\nabla \cdot \overline{(\rho_1 v_1)},$$

$$\mu \nabla^2 \overline{v_2} + \beta \mu \nabla (\nabla \cdot \overline{v_2}) - \overline{\nabla p_2} = \overline{\rho_1 \partial_t v_1} - \overline{\rho_0 (v_1 \cdot \nabla) v_1}.$$

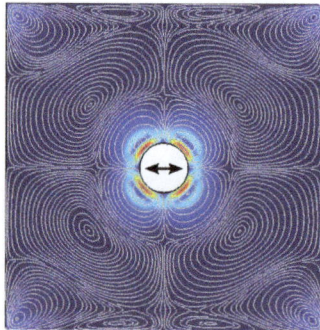

Cross-section of acoustic streaming around a solid cylindrical pillar

In determining the acoustic streaming, the first-order equations are most important. Since

Navier-Stokes equations can only be analytically solved for simple cases, numerical methods are typically used, with the finite element method (FEM) the most common technique. It can be employed to simulate the acoustic streaming phenomena. Figure is one example of acoustic streaming around a solid circular pillar, which is calculated by FEM.

As mentioned, acoustic streaming is driven by mass and force sources originating from the acoustic attenuation. However, these are not the only driven forces for acoustic streaming. The boundary vibration may also contribute, especially to "boundary driven streaming". For these cases, the boundary condition should also be processed by the perturbation approach and be imposed on the two order equations accordingly.

Particle Motion

The motion of a suspended particle whose gravity is balanced by the buoyancy force in an acoustic field is determined by two forces: the acoustic radiation force and Stokes drag force. By applying Newton's law, the motion can be described as:

$$m\frac{du}{dt} = F^{rad} + F^{drag},$$

$$F^{drag} = 6\pi a\mu(v-u).$$

where:

- v is the fluid velocity,
- u is the velocity of particle.

For applications in a static flow, the fluid velocity comes from the acoustic streaming. The magnitude of acoustic streaming depends on the power and frequency of the input and the properties of the fluid media. For typical acoustic-based microdevices, the operating frequency may be from the kHz to the MHz range. The vibration amplitude is in a range of 0.1 nm to 1 μm. Assuming the fluid used is water, the estimated magnitude of acoustic streaming is in the range of 1 μm/s to 1 mm/s. Thus, the acoustic streaming should be smaller than the main flow for most continuous flow applications. The drag force is mainly induced by the main flow in those applications.

Applications

Cell Separation

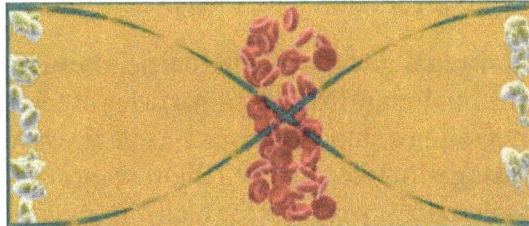

A proposal would use acoustic force to move red blood cells to
a pressure node (center) and lipid cells to antinodes (sides).

Cells with different densities and compression strengths can theoretically be separated with acoustic force. It has been suggested that acoustic tweezers could be used to separate lipid particles from red blood cells. This is a problem during cardiac surgery supported by a heart-lung machine, for which current technologies are insufficient. According to the proposal, acoustic force applied to blood plasma passing through a channel will cause red blood cells to gather in the pressure node in the center and the lipid particles to gather in antinodes at the sides. At the end of the channel, the separated cells and particles exit through separate outlets.

The acoustic method might also be used to separate particles of different sizes. According to the equation of primary acoustic radiation force, larger particles experience larger forces than smaller particles. Shi *et al.* reported using interdigital transducers (IDTs) to generate a standing surface acoustic wave (SSAW) field with pressure nodes in the middle of a microfluidic channel, separating microparticles with different diameters. When introducing a mixture of particles with different sizes from the edge of the channel, larger particles will migrate toward the middle more quickly and be collected at the center outlet. Smaller particles will not be able to migrate to the center outlet before they are collected from the side outlets. This experimental setup has also been used to separate blood components, bacteria, and hydrogel particles.

3D Cell Focusing

Fluorescence-activated cell sorters (FACS) can sort cells by focusing a fluid stream containing the cells, detecting fluorescence from individual cells, and separating the cells of interest from other cells. They have high throughput but are expensive to purchase and maintain, and are bulky with a complex configuration. They also affect cell physiology with high shear pressure, impact forces and electromagnetic forces, which may result in cellular and genetic damage. Acoustic forces are not dangerous to cells, and there has been progress integrating acoustic tweezers with optical/electrical modules for simultaneous cell analysis and sorting, in a smaller and less-expensive machine.

Acoustic tweezers have been developed to achieve 3D focusing of cells/particles in microfluidics. A pair of interdigital transducers (IDTs) are deposited on a piezoelectric substrate, and a microfluidic channel is bonded with the substrate and positioned between the two IDTs. Microparticle solutions are infused into the microfluidic channel by a pressure-driven flow. Once an RF signal is applied to both IDTs, two series of surface acoustic waves (SAW) propagate in opposite directions toward the particle suspension solution inside the microchannel. The constructive interference of the two SAWs results in the formation of a SSAW. Leakage waves in the longitudinal mode are generated inside the channel, causing pressure fluctuations that act laterally on the particles. As a result, the suspended particles inside the channel will be forced toward either the pressure nodes or antinodes, depending on the density and compressibility of the particles and the medium. When the channel width covers only one pressure node (or antinode), the particles will be focused in that node.

In addition to focusing in a horizontal direction, cells/particles can also be focused in the vertical direction. After SSAW is on, the randomly distributed particles are focused into a single file stream in the vertical direction. By integrating a standing surface acoustic wave (SSAW)-based microdevice capable of 3D particle/cell focusing with laser-induced fluorescence (LIF) detection system, acoustic tweezers are developed into a microflow cytometer for high-throughput single cell analysis.

The tunability offered by chirped interdigital transducers renders it capable of precisely sorting cells into a number (e.g., five) of outlet channels in a single step. This is a major advantage over most existing sorting methods, which typically only sort cells into two outlet channels.

Noninvasive Cell Trapping and Patterning

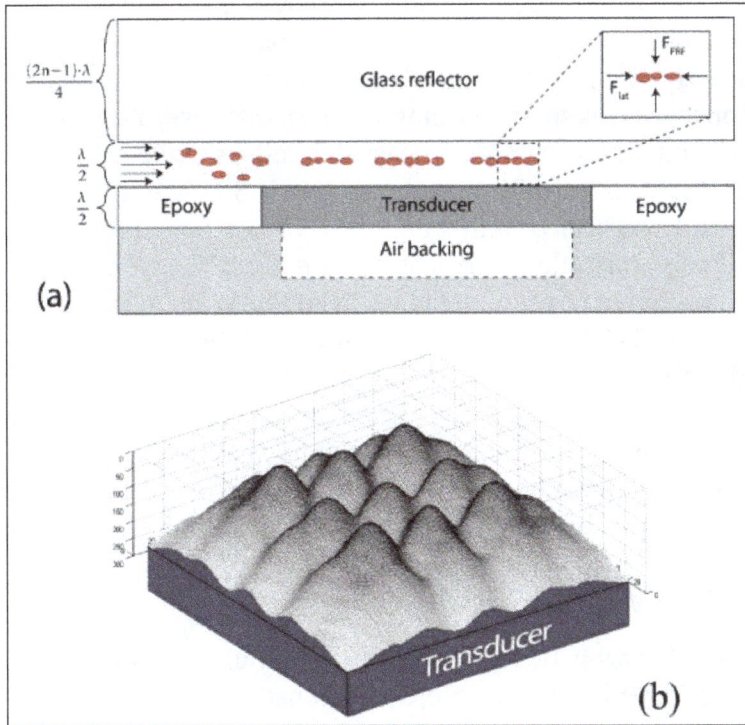

Schematic of a microfluidic device (top) and a 3D graph of trapping node distribution (bottom).

In figure shows a side-view schematic of a microfluidic device. The glass reflector with etched fluidic channels is clamped to the PCB holding the transducer. Cells infused into the chip are trapped in the ultrasonic standing wave formed in the channel. The acoustic forces focus the cells into clusters in the center of the channel as illustrated in the inset. Since the trapping occurs close to the transducer surface, the actual trapping sites are given by the near-field pressure distribution as shown in the 3D image. Cells will be trapped in clusters around the local pressure minima creating different patterns depending on the number of cells trapped. The peaks in the graph correspond to the pressure minima.

Manipulation of Single Cell, Particle or Organism

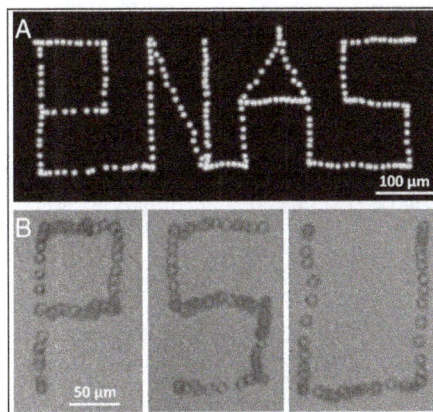

Traces of single cell manipulation

Manipulating single cells is important to many biological studies, such as in controlling the cellular microenvironment and isolating specific cells of interest. Acoustic tweezers have been demonstrated to manipulate each individual cell with micrometer-level resolution. Cells generally have a diameter of 10–20 μm. To meet the resolution requirements of manipulating single cells, short-wavelength acoustic waves should be employed. In this case, a surface acoustic wave (SAW) is preferred to a bulk acoustic wave (BAW), because it allows using shorter-wavelength acoustic waves (normally less than 200 μm). Ding *et al.* reported a SSAW microdevice that is able to manipulate single cells with prescribed paths. Figure records a demonstration that the movement of single cells can be finely controlled with acoustic tweezers. The working principle of the device lies in the controlled movement of pressure nodes in an SSAW field. Ding *et al.* employed chirped interdigital transducers (IDTs) that are able to generate SSAWs with adjustable positions of pressure nodes by changing the input frequency. They also showed that the millimeter-sized microorganism *C. elegan* can be manipulated in the same manner. They also examined cell metabolism and proliferation after acoustic treatment, and found no significant differences compared to the control group, indicating the non-invasive nature of acoustic base manipulation. In addition to using chirped IDTs, phaseshift-based single particle/cell manipulation has also been reported.

Manipulation of Single Biomolecules

Sitters *et al.* have shown that acoustics can be used to manipulate single biomolecules such as DNA and proteins. This method, which the inventors call acoustic force spectroscopy, allows measuring the force response of single molecules. This is achieved by attaching small microspheres to the molecules at one side and attaching them to a surface at the other. By pushing the microspheres away from the surface with a standing acoustic wave the molecules are effectively stretched out.

Manipulation of Organic Nano-materials

Polymer-dispersed liquid crystal (PDLC) displays can be switched from opaque to transparent using acoustic tweezers. A SAW-driven PDLC light shutter has been demonstrated by integrating a cured PDLC film and a pair of interdigital transducers (IDTs) onto a piezoelectric substrate.

Manipulation of Inorganic Nano-materials

Acoustic tweezers provide a simple approach for tuneable nanowire patterning. In this approach, SSAWs are generated by interdigital transducers, which induced a periodic alternating current (AC) electric field on the piezoelectric substrate and consequently patterned metallic nanowires in suspension. The patterns could be deposited onto the substrate after the liquid evaporated. By controlling the distribution of the SSAW field, metallic nanowires are assembled into different patterns including parallel and perpendicular arrays. The spacing of the nanowire arrays could be tuned by controlling the frequency of the surface acoustic waves.

Selective Manipulation

While most acoustic tweezers are able to manipulate a large number of objects collectively, a complementary function is to be able to manipulate a single particle within a cluster without moving adjacent objects. To achieve this goal, the acoustic trap must be localized spacially. A first approach consists in using highly focused acoustic beams. Since many particle of interest are attracted to the nodes of an acoustic field and thus expelled from the focus point, some specific wave structures combining strong focalization but with a minimum of the pressure amplitude at the focal point (surrounded by a ring of intensity to create the trap) are required to trap this type of particle. These specific conditions are met by Bessel beams of topological order larger than zero, also called "acoustical vortices". With this kind of wave structures, the 3D selective manipulation of particles has been demonstrated with an array of transducers driven by programmable electronics.

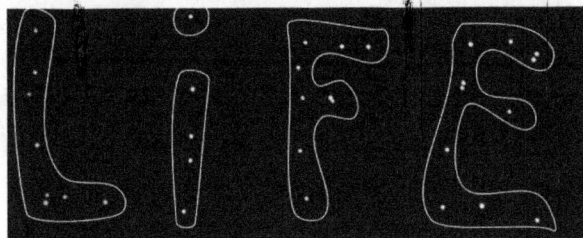

Polystyrene microspheres arranged into a pattern using selective acoustic tweezers.

Compact flat acoustic tweezers based on spiral-shaped interdigital transducers have been proposed as an alternative to this complex array of transducer. This type of device allows patterning dozens of microscopic particles on a microscope slide. The selectivity was nevertheless limited since the acoustic vortex was only focused laterally and hence some spurious secondary rings of weaker could also trap particles. Greater selectivity has been achieved by generating spherically focused acoustical vortices with a flat holographic transducer, combining underlying physical principles of Fresnel lenses in optics, the specificity of Bessel beam topology, and the principles of wave synthesis with IDTs. These latter tweezers generate spherically focused acoustical vortices, and hold potential for 3D manipulation of particles. Alternatively, another approach to localize the acoustic energy relies on the use of nanosecond-scale pulsed fields to generate localized acoustic standing waves.

Applications of Sound

Sonar

SONAR is a technique that uses sound waves to map or locate objects in the surrounding environment. The premise is quite simple: first, emit a cluster of sound waves in the direction of an object. While a few waves will bounce off it, the remaining waves will be reflected back in the direction of the emitter.

If you were to insert one end of a tube into an enormous sea and put an ear to the other, you would definitely look like a loon. However, you would also hear the faint groaning of ships and the singing of various animals far away in the vast depths of the ocean. Leonardo Da Vinci was the first person to perform this ingenious experiment (without the fear of being judged) and discovered this whimsical phenomenon. He had successfully implemented what we now call SONAR.

Sound Navigation and Ranging

Sonar is a technique that uses sound waves to map or locate objects in the surrounding environment. The technique isn't something extravagant that humans have developed in recent years; it has been used by animals such as bats and whales for millions of years.

A bat flying while the sun sets.

The premise is quite simple: first, emit a cluster of sound waves in the direction of an object. While a few waves will bounce off it, the remaining waves will be reflected back in the direction of the emitter. With the knowledge of the speed of sound and the time that passed before the wave was retrieved, an adroit receiver can calculate the object's distance from the emitter.

While Sonar can be implemented in the open air, it is known to be more effective in water. This is because sound waves tend to travel longer distances in water. Owing to Sonar's remarkable range, whales can discern the shape and movement of objects the size of ping-pong balls from 50 feet away. They are known to rely on Sonar even more than sight to forage and track their kin.

A pod of whales.

Active and Passive Sonar

Eventually, humans developed Sonar machines with exponentially superior range and resolution. The simplest of them comprises the combinatorial system of our voice box and ears. It is Sonar that we implement atop mountains and in canyons when we yell at the top of our lungs and eventually hear the echo. However, the LFA Sonar developed by the military emits sound waves that travel thousands of miles. Its sweeping range enables us to cover almost 80% of Earth's oceans by emitting sound waves from only four vantage points.

Despite light's and, for that matter, RADAR's tremendously superior velocity, it is Sonar that is used by NOAA to develop nautical charts, execute seafloor mapping, locate shipwrecks and predict underwater hazards. In fact, Sonar's patent was sanctioned after witnessing the events that led to the Titanic's tragic undoing. Its primary purpose was to identify objects lurking beneath the ocean's surface in order to avoid underwater collisions.

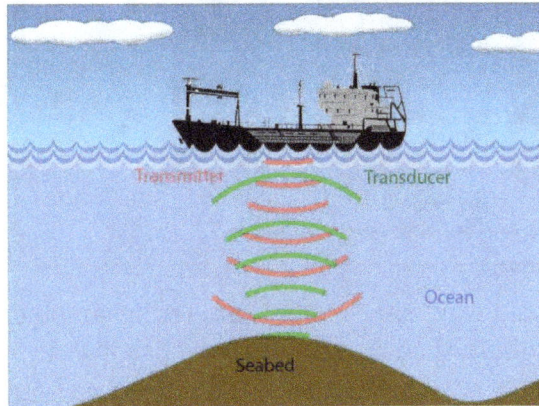

A representation of how ships use SONAR to map seafloors.

Subsequently, the first World War brought major advancements that paved the way for underwater surveillance and warfare submarines. Underwater surveillance implements what is known as passive Sonar — a technique that does not require its own transmitter, as it entails listening to sound waves emitted by other transmitters. This means listening to the sounds made by whales and enemy ships. The tool simply detects the sound waves that travel towards it. The machines, however, cannot determine the locations of these transmitters without the help of other passive listening devices. They work in conjunction to triangulate the location of a transmitter, stealthily, without making their presence felt.

Submarines transmit sound waves and detect objects in their vicinity by measuring the elapsed time between the reception of the echo.

On the other hand, warfare submarines implement active Sonar — a technique that utilizes a receiver as well as a transmitter. This is the technique we most readily associate with Sonar. Submarines transmit sound waves and detect objects in their vicinity by measuring the elapsed time before they receive the echo. Other than merely detecting an object's presence, the gradual rise of superiorly sophisticated tools has also allowed us to identify the shape, size and orientation in exquisite detail.

Compromise between Resolution and Attenuation

The transmitters are mostly piezoelectric materials, materials that wobble and distort when subjected to an electric current. The production of sound from these distortions is analogous to the vibration of a diaphragm in your speaker. Conversely, piezoelectric materials produce an electric current when subjected to distortion, a property that convinced us to simultaneously employ them as receivers.

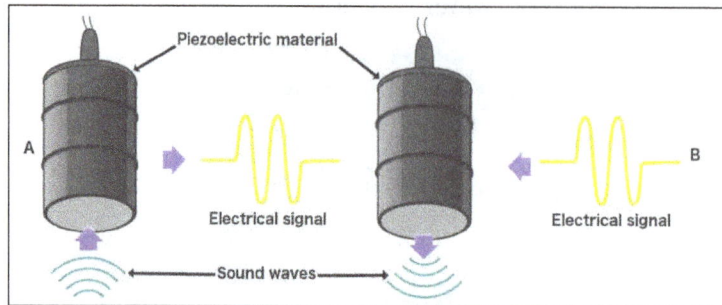

However, because the reflected waves are waves scattered by an object, one can reasonably conclude that their intensity is diminished compared to the original, incident sound waves. The low intensity of the received waves renders images murky or not suitably bright. The quality of an image, therefore, depends not only on the capabilities of the machine, but also the aspects of the object and the terrain in which the mechanism is implemented.

For instance, objects covered with more craggy or irregular surfaces absorb more sound waves than objects covered with regular or smooth surfaces. The propagation of sound waves can also be affected by the temperature of water and the impurities that it fosters. Resolution and range, on the other hand, are characteristics that are intimately linked to the frequency of sound waves.

Sound travels longer distances in water than light or radio waves.

Low-frequency sound waves, those below 20kHz, generate poor resolution, but they boast higher ranges, as they are highly unlikely to be attenuated by obstacles in between. On the contrary, high-frequency sound waves, those with a frequency greater than 100kHz, generate phenomenal resolution, but they are prone to heavy attenuation. A compromise emerges, such that the optimal frequency must be carefully selected in proportion to the size of the desired detail.

Sonar isn't solely used for surveillance or by warfare submarines; it is also used by doctors to detect cysts and cancerous cells, a process which is known as ultrasonography. Doctors infiltrate their patients with sound waves that scatter and ricochet inside the body, enabling them to detect muscles and organs in much greater detail than X-rays would allow.

Sonar devices are also attached to the ends of fishing nets, allowing fishermen to get a rough estimate of the fish caught in the net. Even Batman couldn't resist using Sonar, albeit unethically, to catch the evasive Joker. Despite the scene's unrealistic or far-fetched implementation of Sonar, one realizes how substantial the technology is.

Echolocation

Echolocation is a technique used by bats, dolphins and other animals to determine the location of objects using reflected sound. This allows the animals to move around in pitch darkness, so they can navigate, hunt, identify friends and enemies, and avoid obstacles.

Animals that use Echolocation

The lowland streaked tenrec from Madagascar is one of the more unusual animals that has evolved echolocation.

Bats, whales, dolphins, a few birds like the nocturnal oilbird and some swiftlets, some shrews and the similar tenrec from Madagascar are all known to echolocate. Another possible candidate is the hedgehog, and incredibly some blind people have also developed the ability to echolocate.

Why did Echolocation Evolve in Animals

For dolphins and toothed whales, this technique enables them to see in muddy waters or dark ocean depths, and may even have evolved so that they can chase squid and other deep-diving species.

Echolocation allows bats to fly at night as well as in dark caves. This is a skill they probably developed so they could locate night-flying insects that birds can't find.

It would be impossible for bats to fly around at night without echolocation.

How do Dolphins use Echolocation

Dolphins and whales use echolocation by bouncing high-pitched clicking sounds off underwater objects, similar to shouting and listening for echoes. The sounds are made by squeezing air through nasal passages near the blowhole. These soundwaves then pass into the forehead, where a big blob of fat called the melon focuses them into a beam.

If the echolocating call hits something, the reflected sound is picked up through the animal's lower jaw and passed to its ears. Echolocating sounds are so loud that the ears of dolphins and whales are shielded to protect them. Dolphins and whales use this method to work out an object's distance, direction, speed, density and size.

Schools of dolphins use echolocation to communicate with each other and hunt.

Using echolocation, dolphins can detect an object the size of a golfball about the length of a football pitch away – much further than they can see. By moving its head to aim the sound beam at different parts of a fish, a dolphin can also differentiate between species.

How do Bats use Echolocation

Bats make echolocating sounds in their larynxes and emit them through their mouths. Fortunately, most are too high-pitched for humans to hear – some bats can scream at up to 140 decibels, as loud as a jet engine 30m away.

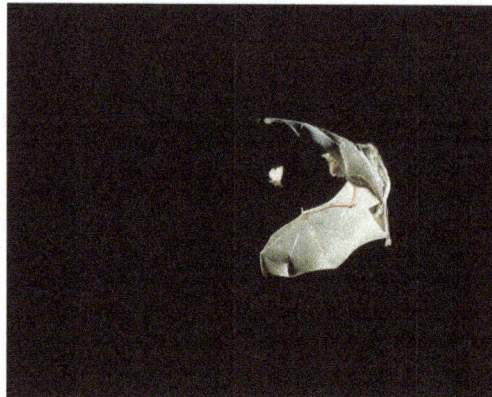

Greater horseshoe bat using echolocation to chase a moth.

Bats can detect an insect up to 5m away, work out its size and hardness, and can also avoid wires as fine as human hairs. As a bat closes in for the kill, it cranks up its calls to pinpoint the prey.

To avoid being deafened by its own calls, a bat turns off its middle ear just before calling, restoring its hearing a split second later to listen for echoes.

How do other Animals use Echolocation

The oilbird is active at night, and some insect-eating swiftlets roost in dark caves, so it makes sense for them to have evolved the ability to echolocate. Both use sharp, audible clicks to navigate through the darkness.

Some nocturnal shrews use ultrasonic squeaks to explore their dark surroundings, and the shrew-like tenrecs of Madagascar echolocate at night using tongue clicks, possibly to find food.

Hedgehogs use ultrasonic whistles, they've got excellent hearing and they live in similar habitats to tenrecs and shrews, but we haven't yet been able to confirm that they echolocate for certain.

Another intriguing possibility is humans – many blind people can find their way around simply by listening to echoes bouncing off surrounding objects, and some expert human echolocators make short high clicks similar to those found in nature.

References

- Categories-of-Waves, Lesson-1, waves, class: physicsclassroom.com, Retrieved 12 April, 2019

- What-is-sonar-definition-active-passive-uses-examples, innovation:scienceabc.com, Retrieved 1 May, 2019

- S. P. Näsholm and S. Holm, "On a Fractional Zener Elastic Wave Equation," Fract. Calc. Appl. Anal. Vol. 16, No 1 (2013), pp. 26-50, DOI: 10.2478/s13540-013--0003-1

- What-is-echolocation, mammals, animal-facts: discoverwildlife.com, Retrieved 25 August, 2019

- Saw, fundamentals-of-low-dimensional-semiconductor-systems, research: sp.phy.cam.ac.uk, Retrieved 19 April, 2019

- Dion, J. L.; Malutta, A.; Cielo, P. (1982). "Ultrasonic Inspection Of Fiber Suspensions". Journal of the Acoustical Society of America. 72(5): 1524–1526. Bibcode:1982ASAJ...72.1524D. Doi:10.1121/1.388688

- Acoustic-noise-causes-rating-systems-and-design-guidelines: comairrotron.com, Retrieved 21 March, 2019

- Difference-between-sound-and-noise, words-language, language: differencebetween.net, Retrieved 20 July, 2019

2
Sub-disciplines of Acoustics

Acoustics can be divided into various sub-disciplines. These include aeroacoustics, archaeo-acoustics, acoustic signal processing, architectural acoustics, bioacoustics, musical acoustics, underwater acoustics and psychoacoustics. The topics elaborated in this chapter will help in gaining a better perspective about these sub-disciplines of acoustics.

Aeroacoustics

The generation of sound by a turbulent flow is the most common physical effect associated with the field of aeroacoustics. The prefix aero means air, however, the field of aeroacoustics is not restricted to flow-induced noise in air. Aeroacoustics is concerned with the general interaction between a background flow and an acoustics field. For example, if you were studying the reflections of sound off a shear layer or how the flow in a muffler affects transmission loss, you would be studying aeroacoustics.

Aeroacoustics models can include effects that alter an acoustic field in the presence of a flow, such as turbulence, local variations in material properties, convection, viscous damping, and much more. Numerically solving aeroacoustic problems falls under the field of computational aeroacoustics (CAA).

Flow-induced Noise

Acoustic noise generated by a flow can be created through different mechanics, but is ultimately due to fluctuations in the flow. These fluctuations will give rise to distributed acoustic sources throughout the flow. Noise is created by local stress fluctuations in the flow (Reynolds stresses, viscous stress effects, and nonisentropic effects all act as quadrupole sources); pressure fluctuations at walls (e.g., dipole sources at solid boundaries); mass and heat fluctuations (e.g., distributed monopole sources); and external fluctuating force fields.

Acoustic waves are only a subset of waves that are created by the flow; vorticity and thermal instability waves are also produced. These particular waves are only convected by the flow, whereas acoustic waves also propagate relative to the flow, at the local speed of sound. Compared to other fluctuations, the energy associated with acoustic waves is typically many orders of magnitude smaller. This means that a direct simulation of the flow-induced noise is very challenging and requires extremely accurate numerical schemes. Another difficulty is that the acoustic sources occur at the length scale of the turbulence in the flow, which then needs to be resolved.

Acoustic Analogies

Acoustic analogy equations are formulations that separate out acoustic fluctuations from flow fluctuations.

The first form of these equations derived from the seminal work of J. Lighthill in the 1950s. His acoustic analogy equation is obtained by rearranging the Navier-Stokes equations, which result in a convected scalar wave equation for the acoustic pressure and a right-hand side that contains all of the flow source terms.

Analogy equations can also be constructed by rearranging the flow equations in other ways, leading to linearized Euler equations or linearized Navier-Stokes equations with right-hand side flow sources. The linearized models have great applicability in many aeroacoustics situations where the flow-induced noise is not important. This is found in fluid-structure interaction (FSI) problems in the frequency domain, when modeling how a background flow affects the acoustics (e.g., reflection and refraction), detailed muffler models, and much more.

Linearized Models of Aeroacoustics

Ideally, aeroacoustic simulations would involve solving the fully compressible Navier-Stokes equations in the time domain. Acoustic pressure waves would then form a subset of the fluid solution. However, this approach is often impractical for real-world applications, due to the computational time and memory resources required.

Instead, like solving for many practical engineering problems, a decoupled two-step multiphysics approach is used:

1. Solve for the fluid flow.

2. Solve for the acoustic perturbations on top of the flow.

Several models exist that include different levels of detail and different aeroacoustic effects. These models are derived by linearizing the dependent variables around a steady-state background flow and making physical assumptions. The dependent variables now represent perturbations, such as acoustic and instability waves, on top of the flow. The classical scalar wave equation, otherwise known as the Helmholtz equation in the frequency domain, is an example of a linearized model with no explicit losses and an isentropic assumption.

Types of Linearized Models

Scalar Wave Equation

In its general form, the scalar wave equation includes the dependency of the speed of sound and density on the background flow (as they can both be spatially varying). It is a good approximation when convective effects are not important (i.e., when the Mach number is less than 0.1) and thermal and viscous losses are also negligible.

Thermoacoustics

Linearized thermoacoustics models involve the dependency of material parameters on the background flow, as well as thermal and viscous losses. They are sufficient models when the Mach number is less than 0.1. An application example is modeling FSI in the frequency domain.

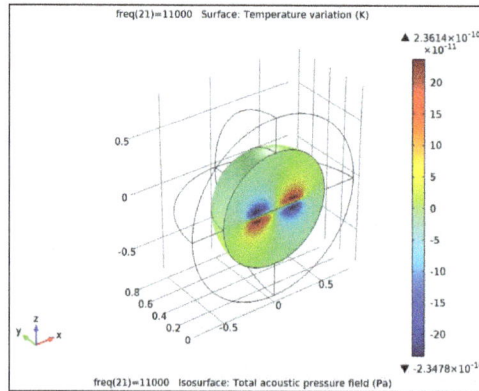

Model of a vibrating micromirror. The plot shows the acoustic variations of the temperature field.

Linearized Potential Flow

Linearized potential flow models involve the interaction between a background potential flow and the acoustic field. Here, no losses are modeled except for possible impedance conditions at walls. Linearized potential flow can be used for modeling sound propagation from jet engines, for example.

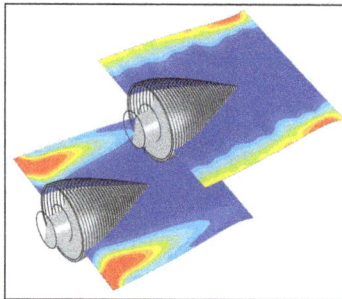

Acoustic pressure distribution on a lined duct wall of a jet engine.

Linearized Euler

Linearized Euler models can be used to compute acoustic variations in density, velocity, and pressure in the presence of a stationary background mean flow that is well approximated by an ideal gas flow. In its general form, this equation supports both acoustic and nonacoustic waves. The latter are so-called instability waves (vorticity and entropy waves).

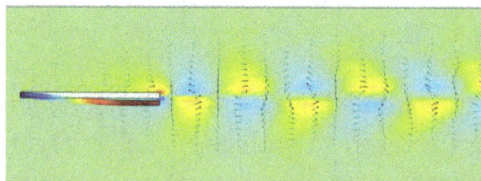

Visualizing the vibrations of a plate in a 2D viscous parallel plate flow.
Modeled with the linearized Navier-Stokes equations.

Linearized Navier-stokes

Linearized Navier-Stokes equations are used to compute the acoustic variations in pressure, velocity, and temperature in the presence of any stationary isothermal or nonisothermal background mean flow. These equations include viscous losses, thermal conduction, and heat generated by viscous dissipation, if relevant.

Acoustic Perturbation Equations

There are many variations and modifications of the aforementioned linearized models for modeling aeroacoustics problems. One such example is the acoustic perturbation equations. Here, non-acoustic propagating modes have been filtered out by modifying the governing equation, resulting in a pure acoustic problem.

Archaeoacoustics

Archaeoacoustics is the use of acoustical study as a methodological approach within the field of archaeology. Archaeoacoustics examines the acoustics of archaeological sites and artifacts. It is an interdisciplinary field that includes archaeology, ethnomusicology, acoustics and digital modelling, and is part of the wider field of music archaeology, with a particular interest in prehistoric music. Since many cultures explored through archaeology were focused on the oral and therefore the aural, researchers believe that studying the sonic nature of archaeological sites and artifacts may reveal new information on the civilizations scrutinized.

Disciplinary Methodology

Damian Murphy of the University of York has studied measurement techniques in acoustic archaeology.

Ancient Sites

Stonehenge in 2007.

Aaron Watson undertook work on the acoustics of numerous archaeological sites, including that of Stonehenge, investigated numerous chamber tombs and other stone circles. Rupert Till

(Huddersfield) and Bruno Fazenda (Salford) also explored Stonehenge's acoustics. In the October 2011 edition of the *Journal of the Acoustical Society of America*, Steven Waller argued that acoustics interference patterns were used to design the blue print of Stonehenge.

Miriam Kolar and colleagues (Stanford) studied various spatial and perceptual attributes of Chavín de Huantar. They identified within the site held the same resonance produced by pututu shells (also used as instruments in the Chavín culture).

Chichen Itza in 2009.

Scientific research led since 1998 suggests that the Kukulkan pyramid in Chichen Itza mimics the chirping sound of the quetzal bird when humans clap their hands around it. The researchers argue that this phenomenon is not accidental, that the builders of this pyramid felt divinely rewarded by the echoing effect of this structure. Technically, the clapping noise rings out and scatters against the temple's high and narrow limestone steps, producing a chirp-like tone that declines in frequency.

Lithophony

Archaeologist Paul Devereux's work has looked at ringing rocks, Avebury and various other subjects, that he details in his book Stone Age Soundtracks. Dr. Ian Cross of University of Cambridge has explored lithoacoustics, the use of stones as musical instruments. Archaeologist Cornelia Kleinitz has studied the sound of a rock gongs in Sudan with Rupert Till and Brenda Baker.

Art and Acoustics

Iegor Reznikoff and Michel Dauvois studied the prehistoric painted caves of France, and found links between the artworks' positioning and acoustic effects. An AHRC project headed by Rupert Till of Huddersfield University, Chris Scarre of Durham University and Bruno Fazenda of Salford University, studies similar relationships in the prehistoric painted caves in northern Spain.

Archaeologists Margarita Díaz-Andreu, Carlos García Benito and Tommaso Mattioli have undertaken work on rock art landscapes in Italy, France and Spain, paying particular attention to echolocation and augmented audibility of distant sounds that is experienced in some rock art sites.

The International Study Group on Music Archaeology (ISGMA), which includes archaeoacoustical work, is a pool of researchers devoted to the field of music archaeology. The study group is hosted at the Orient Department of the German Archaeological Institute Berlin (DAI, Deutsches Archäologisches Institut, Orient-Abteilung) and the Department for Ethnomusicology at the Ethnological Museum of Berlin (Ethnologisches Museum Berlin, SMB SPK, Abteilung Musikethnologie, Medien-Technik und Berliner Phonogramm-Archiv). The ISGMA comprises research methods of musicological and anthropological disciplines (archaeology, organology, acoustics, music iconology, philology, ethnohistory, and ethnomusicology).

The Acoustics and Music of British Prehistory Research Network was funded by the Arts and Humanities Research Council and Engineering and Physical Sciences Research Council, led by Rupert Till and Chris Scarre, as well as Professor Jian Kang of Sheffield University's Department of Architecture. It has a list of researchers working in the field, and links to many other relevant sites. An e-mail list has been discussing the subject since 2002 and was set up as a result of the First Pan-American/Iberian Meeting on Acoustics by Victor Reijs.

Based in the US, the OTS Foundation has conducted several international conferences specifically on Archaeoacoustics, with a focus on the human experience of sound in ancient ritual and ceremonial spaces. The published papers represent a broader multidisciplinary study and include input from the realms of archaeology, architecture, acoustic engineering, rock art, and psycho-acoustics, as well as reports of field work from Gobekli Tepe and Southern Turkey, Malta, and elsewhere around the world.

The European Music Archeology Project is a multi-million euro project to recreate ancient instruments and their sounds, and also the environments in which they would have been played.

Past Interpretations Controversy

An early interpretation of the idea of archaeoacoustics was that it explored acoustic phenomena encoded in ancient artifacts. For instance, the idea that a pot or vase could be "read" like a gramophone record or phonograph cylinder for messages from the past, sounds encoded into the turning clay as the pot was thrown. There is little evidence to support such ideas, and there are few publications claiming that this is the case. In comparison, the more contemporary approach to the field now has many publications and a growing significance. This earlier approach was first raised in the 6 February 1969 issue of *New Scientist* magazine, where it was discussed in David E. H. Jones's light-hearted "Daedalus" column. He wrote:

> "A trowel, like any flat plate, must vibrate in response to sound: thus, drawn over the wet surface by the singing plasterer, it must emboss a gramophone-type recording of his song in the plaster. Once the surface is dry, it may be played back".

—Jones

Jones subsequently received a letter from one Richard G. Woodbridge III who claimed to have already been working on the idea and said that he had sent a paper on the subject to the journal *Nature*. The paper never appeared in *Nature*, but the August 1969 edition of the journal *Proceedings of the IEEE* printed a letter from Woodbridge entitled "Acoustic Recordings from Antiquity". In this communication, the author stated that he wished to call attention to the potential of what he called "Acoustic Archaeology" and to record some early experiments in the field. He then described his experiments with making clay pots and oil paintings from which sound could then be replayed, using a conventional record player cartridge connected directly to a set of headphones. He claimed to have extracted the hum of the potter's wheel from the grooves of a pot, and the word "blue" from an analysis of patch of blue color in a painting.

In 1993, archeology professor Paul Åström and acoustics professor Mendel Kleiner performed similar experiments in Gothenburg, and reported that they could recover some sounds.

An episode of *MythBusters* explored the idea: Killer Cable Snaps, Pottery Record found that while *some* generic acoustic phenomena can be found on pottery, it is unlikely that any discernible sounds (like someone talking) could be recorded on the pots unless ancient people had the technical knowledge to deliberately put the sounds on the artifacts.

In 1902, Charles Sanders Peirce wrote: "Give science only a hundred more centuries of increase in geometrical progression, and she may be expected to find that the sound waves of Aristotle's voice have somehow recorded themselves."

Acoustic Signal Processing

Audio signal processing is a subfield of signal processing that is concerned with the electronic manipulation of audio signals. Audio signals are electronic representations of sound waves—longitudinal waves which travel through air, consisting of compressions and rarefactions. The energy contained in audio signals is typically measured in decibels. As audio signals may be represented in either digital or analog format, processing may occur in either domain. Analog processors operate directly on the electrical signal, while digital processors operate mathematically on its digital representation.

The motivation for audio signal processing began at the beginning of the 20th century with inventions like the telephone, phonograph, and radio that allowed for the transmission and storage of audio signals. Audio processing was necessary for early radio broadcasting, as there were many problems with studio-to-transmitter links. The theory of signal processing and it's application to audio was largely developed at Bell Labs in the mid 20th century. Claude Shannon and Harry Nyquist's early work on communication theory, sampling theory, and Pulse-code modulation laid the foundations for the field. In 1957, Max Mathews became the first person to synthesize audio from a computer, giving birth to computer music.

Analog Signals

An analog audio signal is a continuous signal represented by an electrical voltage or current that is "analogous" to the sound waves in the air. Analog signal processing then involves physically altering the continuous signal by changing the voltage or current or charge via electrical circuits.

Historically, before the advent of widespread digital technology, analog was the only method by which to manipulate a signal. Since that time, as computers and software have become more capable and affordable and digital signal processing has become the method of choice. However, in music applications analog technology is often still desirable as it often produces nonlinear responses that are difficult to replicate with digital filters.

Digital Signals

A digital representation expresses the audio waveform as a sequence of symbols, usually binary numbers. This permits signal processing using digital circuits such as digital signal processors, microprocessors and general-purpose computers. Most modern audio systems use a digital approach as the techniques of digital signal processing are much more powerful and efficient than analog domain signal processing.

Application Areas

Processing methods and application areas include storage, data compression, music information retrieval, speech processing, localization, acoustic detection, transmission, noise cancellation, acoustic fingerprinting, sound recognition, synthesis, and enhancement (e.g. equalization, filtering, level compression, echo and reverb removal or addition, etc).

Audio Broadcasting

Audio signal processing is used when broadcasting audio signals in order to enhance their fidelity or optimize for bandwidth or latency. In this domain, the most important audio processing takes place just before the transmitter. The audio processor here must prevent or minimize overmodulation, compensate for non-linear transmitters (a potential issue with medium wave and shortwave broadcasting), and adjust overall loudness to desired level.

Active Noise Control

Active noise control is a technique designed to reduce unwanted sound. By creating a signal that is identical to the unwanted noise but with the opposite polarity, the two signals cancel out due to destructive interference.

Audio Synthesis

Audio synthesis is the electronic generation of audio signals. A musical instrument that accomplishes this is called a synthesizer. Synthesizers can either imitate sounds or generate new ones. Audio synthesis is also used to generate human speech using speech synthesis.

Audio Effects

Audio effects are systems designed to alter how an audio signal sounds. Unprocessed audio is metaphorically referred to as *dry*, while processed audio is referred to as *wet*.

- Delay or echo: To simulate the effect of reverberation in a large hall or cavern, one or several delayed signals are added to the original signal. To be perceived as echo, the delay has to be of order 35 milliseconds or above. Short of actually playing a sound in the desired environment, the effect of echo can be implemented using either digital or analog methods.

Analog echo effects are implemented using tape delays or bucket-brigade devices. When large numbers of delayed signals are mixed a reverberation effect is produced; The resulting sound has the effect of being presented in a large room.

- Flanger: To create an unusual sound, a delayed signal is added to the original signal with a continuously variable delay (usually smaller than 10 ms). This effect is now done electronically using DSP, but originally the effect was created by playing the same recording on two synchronized tape players, and then mixing the signals together. As long as the machines were synchronized, the mix would sound more-or-less normal, but if the operator placed his finger on the flange of one of the players (hence "flanger"), that machine would slow down and its signal would fall out-of-phase with its partner, producing a phasing comb filter effect. Once the operator took his finger off, the player would speed up until it was back in phase with the master, and as this happened, the phasing effect would appear to slide up the frequency spectrum. This phasing up-and-down the register can be performed rhythmically.

- Phaser: Another way of creating an unusual sound; the signal is split, a portion is filtered with a variable all-pass filter to produce a phase-shift, and then the unfiltered and filtered signals are mixed to produce a comb filter. The phaser effect was originally a simpler implementation of the flanger effect since delays were difficult to implement with analog equipment.

- Chorus: A delayed version of the signal is added to the original signal. The delay has to be short in order not to be perceived as echo, but above 5 ms to be audible. If the delay is too short, it will destructively interfere with the un-delayed signal and create a flanging effect. Often, the delayed signals will be slightly pitch shifted to more realistically convey the effect of multiple voices.

- Equalization: Frequency response is adjusted using audio filter(s) to produce desired spectral characteristics. Frequency ranges can be emphasized or attenuated using low-pass, high-pass, band-pass or band-stop filters. Moderate use of equalization can be used to fine-tune the tonal quality of a recording; extreme use of equalization, such as heavily cutting a certain frequency can create more unusual effects. Band-pass filtering of voice can simulate the effect of a telephone because telephones use band-pass filters.

- Overdrive effects such as the use of a fuzz box can be used to produce distorted sounds, such as for imitating robotic voices or to simulate distorted radiotelephone traffic (e.g., the radio chatter between starfighter pilots in the science fiction film Star Wars). The most basic overdrive effect involves clipping the signal when its absolute value exceeds a certain threshold.

- Pitch shift: This effect shifts a signal up or down in pitch. For example, a signal may be shifted an octave up or down. This is usually applied to the entire signal, and not to each note separately. Blending the original signal with shifted duplicate(s) can create harmonies from one voice. Another application of pitch shifting is pitch correction. Here a musical signal is tuned to the correct pitch using digital signal processing techniques. This effect is ubiquitous in karaoke machines and is often used to assist pop singers who sing out of tune. It is also used intentionally for aesthetic effect in such pop songs as Cher's Believe and Madonna's Die Another Day.

- Time stretching: The complement of pitch shift, that is, the process of changing the speed of an audio signal without affecting its pitch.

- Resonators: Emphasize harmonic frequency content on specified frequencies. These may be created from parametric EQs or from delay-based comb-filters.

- Robotic voice effects are used to make an actor's voice sound like a synthesized human voice.

- Modulation: To change the frequency or amplitude of a carrier signal in relation to a pre-defined signal. Ring modulation is an effect made famous by Doctor Who's Daleks and commonly used throughout sci-fi.

- Compression: The reduction of the dynamic range of a sound to avoid unintentional fluctuation in the dynamics. Level compression is not to be confused with audio data compression, where the amount of data is reduced without affecting the amplitude of the sound it represents.

- 3D audio effects: Place sounds outside the stereo basis.

- Reverse echo: A swelling effect created by reversing an audio signal and recording echo and delay while the signal runs in reverse. When played back forward the last echos are heard before the effected sound creating a rush like swell preceding and during playback. Jimmy Page of Led Zeppelin used this effect in the bridge of "Whole Lotta Love".

- Wave field synthesis: A spatial audio rendering technique for the creation of virtual acoustic environments.

Acoustic Ecology

Acoustic ecology, sometimes called ecoacoustics or soundscape studies, is a discipline studying the relationship, mediated through sound, between human beings and their environment. Acoustic ecology studies started in the late 1960s with R. Murray Schafer and his team at Simon Fraser University (Vancouver, British Columbia, Canada) as part of the World Soundscape Project. The original WSP team included Barry Truax and Hildegard Westerkamp, Bruce Davies and Peter Huse, among others. The first study produced by the WSP was titled The Vancouver Soundscape. The interest in this area grew enormously after this pioneer and innovative study and the area of acoustic ecology raised the interest of researchers and artists all over the world. In 1993, the members of the by now large and active international acoustic ecology community formed the World Forum of Acoustic Ecology.

Every three years since the WFAE's founding at Banff, Canada in 1993, an international symposium has taken place. Stockholm, Amsterdam, Devon, Peterborough, and Melbourne followed. In November 2006, the WFAE meeting took place in Hirosaki, Japan. Koli, Finland, was the meeting place of the latest WFAE world conference.

From its roots in the sonic sociology and radio art of Schafer and his colleagues, acoustic ecology has found expression in many different fields. While most have taken some inspiration from Schafer's writings, in recent years there have also been healthy divergences from the initial ideas. Among the expanded expressions of acoustic ecology are increasing attention to the sonic impacts of road and airport construction, widespread networks of "phonographers" exploring the world through sound, the broadening of bioacoustics (the use of sound by animals) to consider the subjective and objective responses of animals to human noise, including increasing use of the idea of "acoustic ecology" in the literature, and a popular in the effects of human noise on animals, with ocean noise capturing the most attention. Acoustic ecology finds expression in many different fields, including niches as unique as historical soundscapes and psychosonography.

Bioacoustics

Noise is generally a by-product of increased urbanization and development. Noise can alter the acoustic environment of aquatic and terrestrial habitats. Bird diversity has shown to decline because of chronic noise levels in cities and along roadways. Some species such as the urban great tits have changed the frequency of their calls to adapt. In terms of evolution, man-made noise is a much more recent phenomenon. Scientific research has shown that it has potential to change behavior, alter physiology and even restructure animal communities.

Architectural Acoustics

Architectural acoustics (also known as room acoustics and building acoustics) is the science and engineering of achieving a good sound within a building and is a branch of acoustical engineering. The first application of modern scientific methods to architectural acoustics was carried out by Wallace Sabine in the Fogg Museum lecture room who then applied his new found knowledge to the design of Symphony Hall, Boston.

Symphony Hall, Birmingham, an example of the application of architectural acoustics.

Architectural acoustics can be about achieving good speech intelligibility in a theatre, restaurant or railway station, enhancing the quality of music in a concert hall or recording studio, or suppressing noise to make offices and homes more productive and pleasant places to work and live in. Architectural acoustic design is usually done by acoustic consultants.

Building Skin Envelope

This science analyzes noise transmission from building exterior envelope to interior and vice versa. The main noise paths are roofs, eaves, walls, windows, door and penetrations. Sufficient control ensures space functionality and is often required based on building use and local municipal codes. An example would be providing a suitable design for a home which is to be constructed close to a high volume roadway, or under the flight path of a major airport, or of the airport itself.

Inter-space Noise Control

The science of limiting and controlling noise transmission from one building space to another to ensure space functionality and speech privacy. The typical sound paths are ceilings, room partitions, acoustic ceiling panels (such as wood dropped ceiling panels), doors, windows, flanking, ducting and other penetrations. Technical solutions depend on the source of the noise and the path of acoustic transmission, for example noise by steps or noise by (air, water) flow vibrations. An example would be providing suitable party wall design in an apartment complex to minimize the mutual disturbance due to noise by residents in adjacent apartments.

Interior Space Acoustics

This is the science of controlling a room's surfaces based on sound absorbing and reflecting properties. Excessive reverberation time, which can be calculated, can lead to poor speech intelligibility.

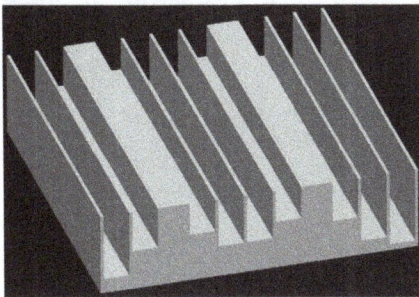

Diffusers which scatter sound are used in some rooms to improve the acoustics

Ceiling of Culture Palace (Tel Aviv) concert hall is covered with perforated metal panels

Sound reflections create standing waves that produce natural resonances that can be heard as a pleasant sensation or an annoying one. Reflective surfaces can be angled and coordinated to provide good coverage of sound for a listener in a concert hall or music recital space. To illustrate this concept consider the difference between a modern large office meeting room or lecture theater and a traditional classroom with all hard surfaces.

Interior building surfaces can be constructed of many different materials and finishes. Ideal acoustical panels are those without a face or finish material that interferes with the acoustical infill or substrate. Fabric covered panels are one way to heighten acoustical absorption. Perforated metal also shows sound absorbing qualities. Finish material is used to cover over the acoustical substrate. Mineral fiber board, or Micore, is a commonly used acoustical substrate. Finish materials often consist of fabric, wood or acoustical tile. Fabric can be wrapped around substrates to create what is referred to as a "pre-fabricated panel" and often provides good noise absorption if laid onto a wall.

An anechoic chamber, using acoustic absorption to create a "dead" space.

Prefabricated panels are limited to the size of the substrate ranging from 2 by 4 feet (0.61 m × 1.22 m) to 4 by 10 feet (1.2 m × 3.0 m). Fabric retained in a wall-mounted perimeter track system, is referred to as "on-site acoustical wall panels". This is constructed by framing the perimeter track into shape, infilling the acoustical substrate and then stretching and tucking the fabric into the perimeter frame system. On-site wall panels can be constructed to accommodate door frames, baseboard, or any other intrusion. Large panels (generally, greater than 50 square feet (4.6 m²)) can be created on walls and ceilings with this method. Wood finishes can consist of punched or routed slots and provide a natural look to the interior space, although acoustical absorption may not be great.

There are four ways to improve workplace acoustics and solve workplace sound problems – the ABCDs.

- A = Absorb (via drapes, carpets, ceiling tiles, etc).

- B = Block (via panels, walls, floors, ceilings and layout).

- C = Cover-up (via sound masking).

- D = Diffuse (cause the sound energy to spread by radiating in many directions).

Mechanical Equipment Noise

Building services noise control is the science of controlling noise produced by:

- ACMV (air conditioning and mechanical ventilation) systems in buildings, termed HVAC in North America.

- Elevators.

- Electrical generators positioned within or attached to a building.

- Any other building service infrastructure component that emits sound.

Inadequate control may lead to elevated sound levels within the space which can be annoying and reduce speech intelligibility. Typical improvements are vibration isolation of mechanical equipment, and sound traps in ductwork. Sound masking can also be created by adjusting HVAC noise to a predetermined level.

Bioacoustics

Bioacoustics is a cross-disciplinary science that combines biology and acoustics. Usually it refers to the investigation of sound production, dispersion and reception in animals (including humans). This involves neurophysiological and anatomical basis of sound production and detection, and relation of acoustic signals to the medium they disperse through. The findings provide clues about the evolution of acoustic mechanisms, and from that, the evolution of animals that employ them.

In underwater acoustics and fisheries acoustics the term is also used to mean the effect of plants and animals on sound propagated underwater, usually in reference to the use of sonar technology for biomass estimation. The study of substrate-borne vibrations used by animals is considered by some a distinct field called biotremology.

For a long time humans have employed animal sounds to recognise and find them. Bioacoustics as a scientific discipline was established by the Slovene biologist Ivan Regen who began systematically to study insect sounds. In 1925 he used a special stridulatory device to play in a duet with an insect. Later, he put a male cricket behind a microphone and female crickets in front of a loudspeaker. The females were not moving towards the male but towards the loudspeaker. Regen's most important contribution to the field apart from realization that insects also detect airborne sounds was the discovery of tympanal organ's function.

Relatively crude electro-mechanical devices available at the time (such as phonographs) allowed only for crude appraisal of signal properties. More accurate measurements were made possible in the second half of the 20th century by advances in electronics and utilization of devices such as oscilloscopes and digital recorders.

The most recent advances in bioacoustics concern the relationships among the animals and their acoustic environment and the impact of anthropogenic noise. Bioacoustic techniques have recently been proposed as a non-destructive method for estimating biodiversity of an area.

Methods

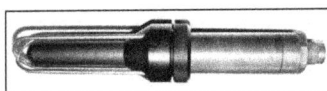

Hydrophone.

Listening is still one of the main methods used in bioacoustical research. Little is known about neurophysiological processes that play a role in production, detection and interpretation of sounds in animals, so animal behaviour and the signals themselves are used for gaining insight into these processes.

Acoustic Signals

An experienced observer can use animal sounds to recognize a "singing" animal species, its location and condition in nature. Investigation of animal sounds also includes signal recording with electronic recording equipment. Due to the wide range of signal properties and media they propagate through, specialized equipment may be required instead of the usual microphone, such as a hydrophone (for underwater sounds), detectors of ultrasound (very high-frequency sounds) or infrasound (very low-frequency sounds), or a laser vibrometer (substrate-borne vibrational signals). Computers are used for storing and analysis of recorded sounds. Specialized sound-editing software is used for describing and sorting signals according to their intensity, frequency, duration and other parameters.

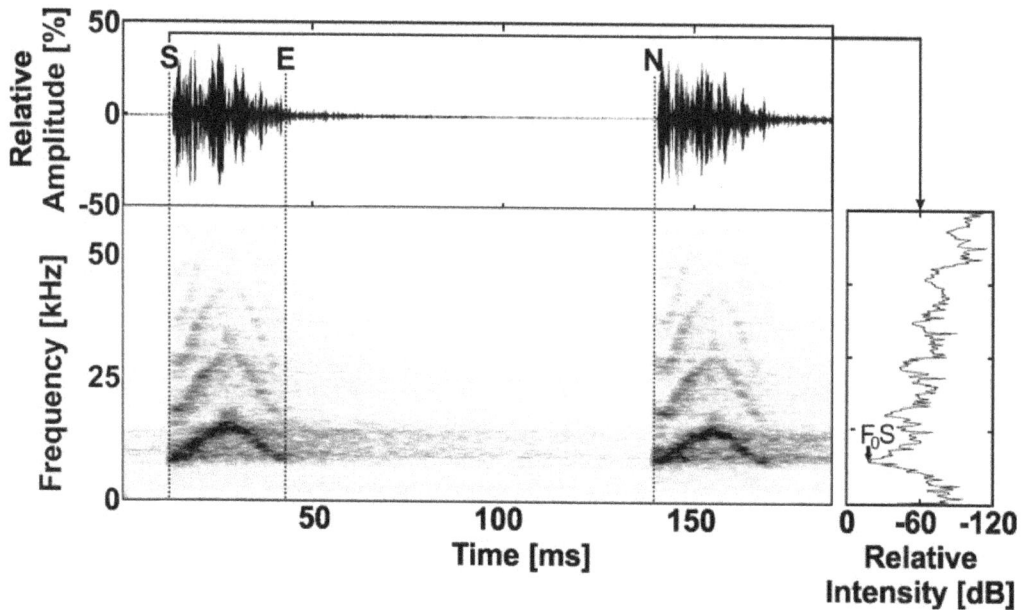

Animal sound collections, managed by museums of natural history and other institutions, are an important tool for systematic investigation of signals. Many effective automated methods involving signal processing, data mining and machine learning techniques have been developed to detect and classify the bioacoustic signals.

Sound Production, Detection and use in Animals

Scientists in the field of bioacoustics are interested in anatomy and neurophysiology of organs involved in sound production and detection, including their shape, muscle action, and activity of neuronal networks involved. Of special interest is coding of signals with action potentials in the latter.

But since the methods used for neurophysiological research are still fairly complex and understanding of relevant processes is incomplete, more trivial methods are also used. Especially useful is observation of behavioural responses to acoustic signals. One such response is phonotaxis – directional movement towards the signal source. By observing response to well defined signals in a controlled environment, we can gain insight into signal function, sensitivity of the hearing apparatus, noise filtering capability, etc.

Biomass Estimation

Biomass estimation is a method of detecting and quantifying fish and other marine organisms using sonar technology. As the sound pulse travels through water it encounters objects that are of different density than the surrounding medium, such as fish, that reflect sound back toward the sound source. These echoes provide information on fish size, location, and abundance. The basic components of the scientific echo sounder hardware function is to transmit the sound, receive, filter and amplify, record, and analyze the echoes. While there are many manufacturers of commercially available "fish-finders," quantitative analysis requires that measurements be made with calibrated echo sounder equipment, having high signal-to-noise ratios.

Animal Sounds

Bergische Crower crowing

European starling singing

Sounds used by animals that fall within the scope of bioacoustics include a wide range of frequencies and media, and are often not "*sound*" in the narrow sense of the word (i.e. compression waves that propagate through air and are detectable by the human ear). Katydid crickets, for example, communicate by sounds with frequencies higher than 100 kHz, far into the ultrasound range. Lower, but still in ultrasound, are sounds used by bats for echolocation. On the other side of the frequency spectrum are low frequency-vibrations, often not detected by hearing organs, but with other, less specialized sense organs. The examples include ground vibrations produced by elephants whose principal frequency component is around 15 Hz, and low- to medium-frequency substrate-borne vibrations used by most insect orders. Many animal sounds, however, do fall within the frequency range detectable by a human ear, between 20 and 20,000 Hz. Mechanisms for sound production and detection are just as diverse as the signals themselves.

Musical Acoustics

Musical acoustics or music acoustics is a branch of acoustics concerned with researching and describing the physics of music – how sounds are employed to make music. Examples of areas of study are the function of musical instruments, the human voice (the physics of speech and singing), computer analysis of melody, and in the clinical use of music in music therapy.

Methods and Fields of Study

- The physics of musical instruments.

- Frequency range of music.

- Fourier analysis.

- Computer analysis of musical structure.

- Synthesis of musical sounds.

- Music cognition, based on physics (also known as psychoacoustics).

Physical Aspects

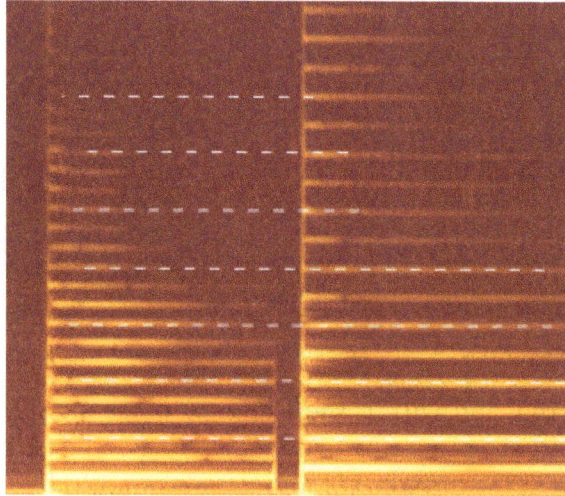

A spectrogram of a violin playing a note and then a perfect fifth above it.
The shared partials are highlighted by the white dashes.

Whenever two different pitches are played at the same time, their sound waves interact with each other – the highs and lows in the air pressure reinforce each other to produce a different sound wave. Any repeating sound wave that is not a sine wave can be modeled by many different sine waves of the appropriate frequencies and amplitudes (a frequency spectrum). In humans the hearing apparatus (composed of the ears and brain) can usually isolate these tones and hear them distinctly. When two or more tones are played at once, a variation of air pressure at the ear "contains" the pitches of each, and the ear and brain isolate and decode them into distinct tones.

When the original sound sources are perfectly periodic, the note consists of several related sine waves (which mathematically add to each other) called the fundamental and the harmonics, partials, or overtones. The sounds have harmonic frequency spectra. The lowest frequency present is the fundamental, and is the frequency at which the entire wave vibrates. The overtones vibrate faster than the fundamental, but must vibrate at integer multiples of the fundamental frequency for the total wave to be exactly the same each cycle. Real instruments are close to periodic, but the frequencies of the overtones are slightly imperfect, so the shape of the wave changes slightly over time.

Subjective Aspects

Variations in air pressure against the ear drum, and the subsequent physical and neurological processing and interpretation, give rise to the subjective experience called sound. Most sound that people recognize as musical is dominated by periodic or regular vibrations rather than non-periodic ones; that is, musical sounds typically have a definite pitch. The transmission of these variations through air is via a sound wave. In a very simple case, the sound of a sine wave, which is considered the most basic model of a sound waveform, causes the air pressure to increase and decrease in a regular fashion, and is heard as a very pure tone. Pure tones can be produced by tuning forks or whistling. The rate at which the air pressure oscillates is the frequency of the tone, which is measured in oscillations per second, called hertz. Frequency is the primary determinant of the perceived pitch. Frequency of musical instruments can change with altitude due to changes in air pressure.

Harmonics, Partials and Overtones

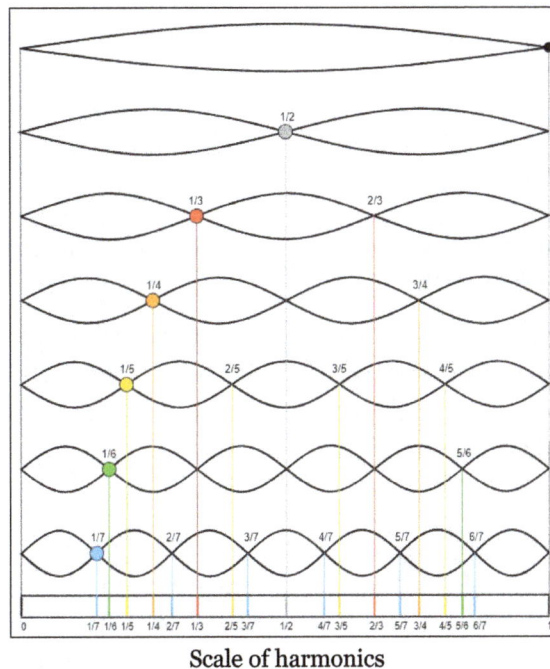

Scale of harmonics

The fundamental is the frequency at which the entire wave vibrates. Overtones are other sinusoidal components present at frequencies above the fundamental. All of the frequency components that make up the total waveform, including the fundamental and the overtones, are called partials. Together they form the harmonic series.

Overtones that are perfect integer multiples of the fundamental are called harmonics. When an overtone is near to being harmonic, but not exact, it is sometimes called a harmonic partial, although they are often referred to simply as harmonics. Sometimes overtones are created that are not anywhere near a harmonic, and are just called partials or inharmonic overtones.

The fundamental frequency is considered the first harmonic and the first partial. The numbering of the partials and harmonics is then usually the same; the second partial is the second harmonic, etc. But if there are inharmonic partials, the numbering no longer coincides. Overtones are numbered as they appear above the fundamental. So strictly speaking, the first overtone is the second partial (and usually the second harmonic). As this can result in confusion, only harmonics are usually referred to by their numbers, and overtones and partials are described by their relationships to those harmonics.

Harmonics and Non-linearities

When a periodic wave is composed of a fundamental and only odd harmonics (f, 3f, 5f, 7f, ...), the summed wave is half-wave symmetric; it can be inverted and phase shifted and be exactly the same. If the wave has any even harmonics (of, 2f, 4f, 6f, ...), it is asymmetrical; the top half is not a mirror image of the bottom.

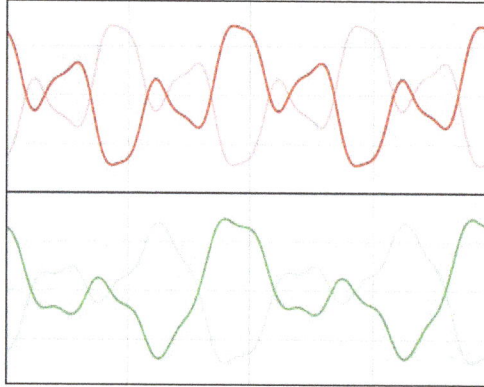

A symmetric and asymmetric waveform. The red (upper) wave contains only the fundamental and odd harmonics; the green (lower) wave contains the fundamental and even harmonics.

Conversely, a system that changes the shape of the wave (beyond simple scaling or shifting) creates additional harmonics (harmonic distortion). This is called a *non-linear system*. If it affects the wave symmetrically, the harmonics produced are all odd. If it affects the harmonics asymmetrically, at least one even harmonic is produced (and probably also odd harmonics).

Harmony

If two notes are simultaneously played, with frequency ratios that are simple fractions (e.g. 2/1, 3/2 or 5/4), the composite wave is still periodic, with a short period—and the combination sounds consonant. For instance, a note vibrating at 200 Hz and a note vibrating at 300 Hz (a perfect fifth, or 3/2 ratio, above 200 Hz) add together to make a wave that repeats at 100 Hz: every 1/100 of a second, the 300 Hz wave repeats three times and the 200 Hz wave repeats twice. Note that the total wave repeats at 100 Hz, but there is no actual 100 Hz sinusoidal component.

Additionally, the two notes have many of the same partials. For instance, a note with a fundamental frequency of 200 Hz has harmonics at: (200,) 400, 600, 800, 1000, 1200, ... A note with fundamental frequency of 300 Hz has harmonics at: (300,) 600, 900, 1200, 1500, ... The two notes share harmonics at 600 and 1200 Hz, and more coincide further up the series.

The combination of composite waves with short fundamental frequencies and shared or closely related partials is what causes the sensation of harmony. When two frequencies are near to a simple fraction, but not exact, the composite wave cycles slowly enough to hear the cancellation of the waves as a steady pulsing instead of a tone. This is called beating, and is considered unpleasant, or dissonant.

The frequency of beating is calculated as the difference between the frequencies of the two notes. For the example above, |200 Hz - 300 Hz| = 100 Hz. As another example, a combination of 3425 Hz and 3426 Hz would beat once per second (|3425 Hz - 3426 Hz| = 1 Hz). This follows from modulation theory.

The difference between consonance and dissonance is not clearly defined, but the higher the beat frequency, the more likely the interval is dissonant. Helmholtz proposed that maximum dissonance would arise between two pure tones when the beat rate is roughly 35 Hz.

Scales

The material of a musical composition is usually taken from a collection of pitches known as a scale. Because most people cannot adequately determine absolute frequencies, the identity of a scale lies in the ratios of frequencies between its tones (known as intervals).

The diatonic scale appears in writing throughout history, consisting of seven tones in each octave. In just intonation the diatonic scale may be easily constructed using the three simplest intervals within the octave, the perfect fifth (3/2), perfect fourth (4/3), and the major third (5/4). As forms of the fifth and third are naturally present in the overtone series of harmonic resonators, this is a very simple process.

The following table shows the ratios between the frequencies of all the notes of the just major scale and the fixed frequency of the first note of the scale.

C	D	E	F	G	A	B	C
1	9/8	5/4	4/3	3/2	5/3	15/8	2

There are other scales available through just intonation, for example the minor scale. Scales that do not adhere to just intonation, and instead have their intervals adjusted to meet other needs are called *temperaments*, of which equal temperament is the most used. Temperaments, though they obscure the acoustical purity of just intervals, often have desirable properties, such as a closed circle of fifths.

Underwater Acoustics

Underwater acoustics is the study of the propagation of sound in water and the interaction of the mechanical waves that constitute sound with the water, its contents and its boundaries. The water may be in the ocean, a lake, a river or a tank. Typical frequencies associated with underwater acoustics are between 10 Hz and 1 MHz. The propagation of sound in the ocean at frequencies lower than 10 Hz is usually not possible without penetrating deep into the seabed, whereas frequencies above 1 MHz are rarely used because they are absorbed very quickly. Underwater acoustics is sometimes known as hydroacoustics.

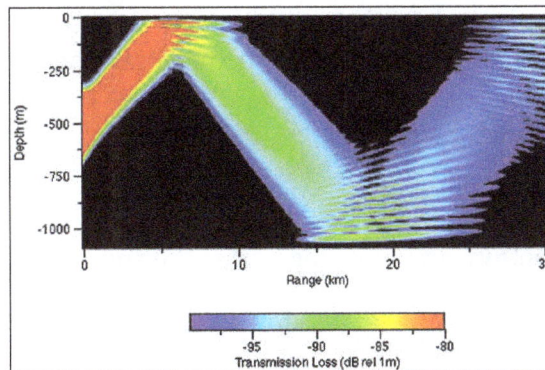

Output of a computer model of underwater acoustic propagation in a simplified ocean environment.

The field of underwater acoustics is closely related to a number of other fields of acoustic study, including sonar, transduction, acoustic signal processing, acoustical oceanography, bioacoustics, and physical acoustics.

Sound Waves in Water, Bottom of Sea

A sound wave propagating underwater consists of alternating compressions and rarefactions of the water. These compressions and rarefactions are detected by a receiver, such as the human ear or a hydrophone, as changes in pressure. These waves may be man-made or naturally generated.

Speed of Sound, Density and Impedance

The speed of sound c (i.e., the longitudinal motion of wavefronts) is related to frequency and f wavelength λ of a wave by $c = f \cdot \lambda$.

This is different from the particle velocity u, which refers to the motion of molecules in the medium due to the sound, and relates the plane wave pressure p to the fluid density ρ and sound speed c by $p = c \cdot u \cdot \rho$.

The product of c and ρ from the above formula is known as the characteristic acoustic impedance. The acoustic power (energy per second) crossing unit area is known as the intensity of the wave and for a plane wave the average intensity is given by $I = q^2 / (\rho c)$, where q is the root mean square acoustic pressure.

At 1 kHz, the wavelength in water is about 1.5 m. Sometimes the term "sound velocity" is used but this is incorrect as the quantity is a scalar.

The large impedance contrast between air and water (the ratio is about 3600) and the scale of surface roughness means that the sea surface behaves as an almost perfect reflector of sound at frequencies below 1 kHz. Sound speed in water exceeds that in air by a factor of 4.4 and the density ratio is about 820.

Absorption of Sound

Absorption of low frequency sound is weak. The main cause of sound attenuation in fresh water, and at high frequency in sea water (above 100 kHz) is viscosity. Important additional contributions at lower frequency in seawater are associated with the ionic relaxation of boric acid (up to c. 10 kHz) and magnesium sulfate (c. 10 kHz-100 kHz).

Sound may be absorbed by losses at the fluid boundaries. Near the surface of the sea losses can occur in a bubble layer or in ice, while at the bottom sound can penetrate into the sediment and be absorbed.

Sound Reflection and Scattering

Boundary Interactions

Both the water surface and bottom are reflecting and scattering boundaries.

Surface

For many purposes the sea-air surface can be thought of as a perfect reflector. The impedance contrast is so great that little energy is able to cross this boundary. Acoustic pressure waves reflected from the sea surface experience a reversal in phase, often stated as either a "pi phase change" or a "180 deg phase change". This is represented mathematically by assigning a reflection coefficient of minus 1 instead of plus one to the sea surface.

At high frequency (above about 1 kHz) or when the sea is rough, some of the incident sound is scattered, and this is taken into account by assigning a reflection coefficient whose magnitude is less than one. For example, close to normal incidence, the reflection coefficient becomes $R = -e^{-2k^2h^2\sin^2 A}$, where h is the rms wave height.

A further complication is the presence of wind generated bubbles or fish close to the sea surface. The bubbles can also form plumes that absorb some of the incident and scattered sound, and scatter some of the sound themselves.

Seabed

The acoustic impedance mismatch between water and the bottom is generally much less than at the surface and is more complex. It depends on the bottom material types and depth of the layers. Theories have been developed for predicting the sound propagation in the bottom in this case, for example by Biot and by Buckingham.

Target

The reflection of sound at a target whose dimensions are large compared with the acoustic wavelength depends on its size and shape as well as the impedance of the target relative to that of water. Formulae have been developed for the target strength of various simple shapes as a function of angle of sound incidence. More complex shapes may be approximated by combining these simple ones.

Propagation of Sound

Underwater acoustic propagation depends on many factors. The direction of sound propagation is determined by the sound speed gradients in the water. This is an important thing that happens in water, because the speed of sound travel in water with velocity regular. In the sea the vertical gradients are generally much larger than the horizontal ones. Combining this with a tendency towards increasing sound speed at increasing depth, due to the increasing pressure in the deep sea, causes a reversal of the sound speed gradient in the thermocline, creating an efficient waveguide at the depth, corresponding to the minimum sound speed. The sound speed profile may cause regions of low sound intensity called "Shadow Zones", and regions of high intensity called "Caustics". These may be found by ray tracing methods.

At equator and temperate latitudes in the ocean, the surface temperature is high enough to reverse the pressure effect, such that a sound speed minimum occurs at depth of a few hundred metres. The presence of this minimum creates a special channel known as Deep Sound Channel, previously known as the SOFAR (sound fixing and ranging) channel, permitting guided propagation of underwater sound for thousands of kilometres without interaction with the sea surface or the seabed. Another phenomenon in the deep sea is the formation of sound focusing areas, known as Convergence Zones. In this case sound is refracted downward from a near-surface source and then back up again. The horizontal distance from the source at which this occurs depends on the positive and negative sound speed gradients. A surface duct can also occur in both deep and moderately shallow water when there is upward refraction, for example due to cold surface temperatures. Propagation is by repeated sound bounces off the surface.

In general, as sound propagates underwater there is a reduction in the sound intensity over increasing ranges, though in some circumstances a gain can be obtained due to focusing. *Propagation loss* (sometimes referred to as *transmission loss*) is a quantitative measure of the reduction in sound intensity between two points, normally the sound source and a distant receiver. If I_s is the far field intensity of the source referred to a point 1 m from its acoustic centre and I_r is the intensity at the receiver, then the propagation loss is given by I_r. In this equation I_r is not the true acoustic intensity at the receiver, which is a vector quantity, but a scalar equal to the equivalent plane wave intensity (EPWI) of the sound field. The EPWI is defined as the magnitude of the intensity of a plane wave of the same RMS pressure as the true acoustic field. At short range the propagation loss is dominated by spreading while at long range it is dominated by absorption and scattering losses.

An alternative definition is possible in terms of pressure instead of intensity, giving $PL = 20\log(p_s / p_r)$, where p_s is the RMS acoustic pressure in the far-field of the projector, scaled to a standard distance of 1 m, and p_r is the RMS pressure at the receiver position.

These two definitions are not exactly equivalent because the characteristic impedance at the receiver may be different from that at the source. Because of this, the use of the intensity definition leads to a different sonar equation to the definition based on a pressure ratio. If the source and receiver are both in water, the difference is small.

Propagation Modeling

The propagation of sound through water is described by the wave equation, with appropriate boundary conditions. A number of models have been developed to simplify propagation calculations. These models include ray theory, normal mode solutions, and parabolic equation simplifications of the wave equation. Each set of solutions is generally valid and computationally efficient in a limited frequency and range regime, and may involve other limits as well. Ray theory is more appropriate at short range and high frequency, while the other solutions function better at long range and low frequency. Various empirical and analytical formulae have also been derived from measurements that are useful approximations.

Reverberation

Transient sounds result in a decaying background that can be of much larger duration than the original transient signal. The cause of this background, known as reverberation, is partly due to scattering from rough boundaries and partly due to scattering from fish and other biota. For an acoustic signal to be detected easily, it must exceed the reverberation level as well as the background noise level.

Doppler Shift

If an underwater object is moving relative to an underwater receiver, the frequency of the received sound is different from that of the sound radiated (or reflected) by the object. This change in frequency is known as a Doppler shift. The shift can be easily observed in active sonar systems, particularly narrow-band ones, because the transmitter frequency is known, and the relative motion between sonar and object can be calculated. Sometimes the frequency of the radiated noise (a tonal) may also be known, in which case the same calculation can be done for passive sonar. For active systems the change in frequency is 0.69 Hz per knot per kHz and half this for passive systems as propagation is only one way. The shift corresponds to an increase in frequency for an approaching target.

Intensity Fluctuations

Though acoustic propagation modelling generally predicts a constant received sound level, in practice there are both temporal and spatial fluctuations. These may be due to both small and large scale environmental phenomena. These can include sound speed profile fine structure and frontal zones as well as internal waves. Because in general there are multiple propagation paths between a source and receiver, small phase changes in the interference pattern between these paths can lead to large fluctuations in sound intensity.

Non-linearity

In water, especially with air bubbles, the change in density due to a change in pressure is not exactly linearly proportional. As a consequence for a sinusoidal wave input additional harmonic and subharmonic frequencies are generated. When two sinusoidal waves are input, sum and difference frequencies are generated. The conversion process is greater at high source levels than small ones. Because of the non-linearity there is a dependence of sound speed on the pressure amplitude so that large changes travel faster than small ones. Thus a sinusoidal waveform gradually becomes a sawtooth one with a steep rise and a gradual tail. Use is made of this phenomenon in parametric sonar and theories have been developed to account for this, e.g. by Westerfield.

Measurements

Sound in water is measured using a hydrophone, which is the underwater equivalent of a microphone. A hydrophone measures pressure fluctuations, and these are usually converted to sound pressure level (SPL), which is a logarithmic measure of the mean square acoustic pressure.

Measurements are usually reported in one of three forms:

- RMS acoustic pressure in micropascals (or dB re 1 µPa).

- RMS acoustic pressure in a specified bandwidth, usually octaves or thirds of octave (dB re 1 µPa).

- Spectral density (mean square pressure per unit bandwidth) in micropascals-squared per hertz (dB re 1 µPa^2/Hz).

The scale for acoustic pressure in water differs from that used for sound in air. In air the reference pressure is 20 µPa rather than 1 µPa. For the same numerical value of SPL, the intensity of a plane wave (power per unit area, proportional to mean square sound pressure divided by acoustic impedance) in air is about $20^2 \times 3600 = 1\,440\,000$ times higher than in water. Similarly, the intensity is about the same if the SPL is 61.6 dB higher in the water.

Sound Speed

Approximate values for fresh water and seawater, respectively, at atmospheric pressure are 1450 and 1500 m/s for the sound speed, and 1000 and 1030 kg/m³ for the density. The speed of sound in water increases with increasing pressure, temperature and salinity. The maximum speed in pure water under atmospheric pressure is attained at about 74 °C; sound travels slower in hotter water after that point; the maximum increases with pressure.

Absorption

Many measurements have been made of sound absorption in lakes and the ocean.

Ambient Noise

Measurement of acoustic signals are possible if their amplitude exceeds a minimum threshold, determined partly by the signal processing used and partly by the level of background noise. Ambient noise is that part of the received noise that is independent of the source, receiver and platform characteristics. This it excludes reverberation and towing noise for example.

The background noise present in the ocean, or ambient noise, has many different sources and varies with location and frequency. At the lowest frequencies, from about 0.1 Hz to 10 Hz, ocean turbulence and microseisms are the primary contributors to the noise background. Typical noise spectrum levels decrease with increasing frequency from about 140 dB re 1 µPa²/Hz at 1 Hz to about 30 dB re 1 µPa²/Hz at 100 kHz. Distant ship traffic is one of the dominant noise sources in most areas for frequencies of around 100 Hz, while wind-induced surface noise is the main source between 1 kHz and 30 kHz. At very high frequencies, above 100 kHz, thermal noise of water molecules begins to dominate. The thermal noise spectral level at 100 kHz is 25 dB re 1 µPa²/Hz. The spectral density of thermal noise increases by 20 dB per decade (approximately 6 dB per octave).

Transient sound sources also contribute to ambient noise. These can include intermittent geological activity, such as earthquakes and underwater volcanoes, rainfall on the surface, and biological activity. Biological sources include cetaceans (especially blue, fin and sperm whales), certain types of fish, and snapping shrimp.

Rain can produce high levels of ambient noise. However the numerical relationship between rain rate and ambient noise level is difficult to determine because measurement of rain rate is problematic at sea.

Reverberation

Many measurements have been made of sea surface, bottom and volume reverberation. Empirical models have sometimes been derived from these. A commonly used expression for the band 0.4 to 6.4 kHz is that by Chapman and Harris. It is found that a sinusoidal waveform is spread in frequency due to the surface motion. For bottom reverberation a Lambert's Law is found often to apply approximately, Volume reverberation is usually found to occur mainly in layers, which change depth with the time of day, The under-surface of ice can produce strong reverberation when it is rough.

Bottom Loss

Bottom loss has been measured as a function of grazing angle for many frequencies in various locations, for example those by the US Marine Geophysical Survey. The loss depends on the sound speed in the bottom (which is affected by gradients and layering) and by roughness. Graphs have been produced for the loss to be expected in particular circumstances. In shallow water bottom loss often has the dominant impact on long range propagation. At low frequencies sound can propagate through the sediment then back into the water.

Underwater Hearing

Comparison with Airborne Sound Levels

As with airborne sound, sound pressure level underwater is usually reported in units of decibels, but there are some important differences that make it difficult (and often inappropriate) to compare SPL in water with SPL in air. These differences include:

- Difference in reference pressure: 1 µPa (one micropascal, or one millionth of a pascal) instead of 20 µPa.

- Difference in interpretation: There are two schools of thought, one maintaining that pressures should be compared directly, and the other that one should first convert to the intensity of an equivalent plane wave.

- Difference in hearing sensitivity: Any comparison with (A-weighted) sound in air needs to take into account the differences in hearing sensitivity, either of a human diver or other animal.

Human Hearing

Hearing Sensitivity

The lowest audible SPL for a human diver with normal hearing is about 67 dB re 1 µPa, with greatest sensitivity occurring at frequencies around 1 kHz. This corresponds to a sound intensity 5.4 dB, or 3.5 times, higher than the threshold in air.

Safety Thresholds

High levels of underwater sound create a potential hazard to human divers. Guidelines for exposure of human divers to underwater sound are reported by the SOLMAR project of the NATO Undersea Research Centre. Human divers exposed to SPL above 154 dB re 1 µPa in the frequency range 0.6 to 2.5 kHz are reported to experience changes in their heart rate or breathing frequency. Diver aversion to low frequency sound is dependent upon sound pressure level and center frequency.

Other Species

Aquatic Mammals

Dolphins and other toothed whales are known for their acute hearing sensitivity, especially in the frequency range 5 to 50 kHz. Several species have hearing thresholds between 30 and 50 dB re 1 µPa in this frequency range. For example, the hearing threshold of the killer whale occurs at an RMS acoustic pressure of 0.02 mPa (and frequency 15 kHz), corresponding to an SPL threshold of 26 dB re 1 μPa.

High levels of underwater sound create a potential hazard to marine and amphibious animals. The effects of exposure to underwater noise are reviewed by Southall et al.

Fish

The hearing sensitivity of fish is reviewed by Ladich and Fay. The hearing threshold of the soldier fish, is 0.32 mPa (50 dB re 1 μPa) at 1.3 kHz, whereas the lobster has a hearing threshold of 1.3 Pa at 70 Hz (122 dB re 1 μPa). The effects of exposure to underwater noise are reviewed by Popper et al.

Applications of Underwater Acoustics

Sonar

Sonar is the name given to the acoustic equivalent of radar. Pulses of sound are used to probe the sea, and the echoes are then processed to extract information about the sea, its boundaries and submerged objects. An alternative use, known as *passive sonar*, attempts to do the same by listening to the sounds radiated by underwater objects.

Underwater Communication

The need for underwater acoustic telemetry exists in applications such as data harvesting for environmental monitoring, communication with and between manned and unmanned underwater vehicles, transmission of diver speech, etc. A related application is underwater remote control, in which acoustic telemetry is used to remotely actuate a switch or trigger an event. A prominent example of underwater remote control are acoustic releases, devices that are used to return sea floor deployed instrument packages or other payloads to the surface per remote command at the end of a deployment. Acoustic communications form an active field of research with significant challenges to overcome, especially in horizontal, shallow-water channels. Compared with radio telecommunications, the available bandwidth is reduced by several orders of magnitude. Moreover, the low speed of sound causes multipath propagation to stretch over time delay intervals of tens or hundreds of milliseconds, as well as significant Doppler shifts and spreading. Often acoustic communication systems are not limited by noise, but by reverberation and time variability beyond the capability of receiver algorithms. The fidelity of underwater communication links can be greatly improved by the use of hydrophone arrays, which allow processing techniques such as adaptive beamforming and diversity combining.

Underwater Navigation and Tracking

Underwater navigation and tracking is a common requirement for exploration and work by divers, ROV, autonomous underwater vehicles (AUV), manned submersibles and submarines alike. Unlike most radio signals which are quickly absorbed, sound propagates far underwater and at a rate that can be precisely measured or estimated. It can thus be used to measure distances between a tracked target and one or multiple reference of *baseline stations* precisely, and triangulate the position of the target, sometimes with centimeter accuracy. Starting in the 1960s, this has given rise to underwater acoustic positioning systems which are now widely used.

Seismic Exploration

Seismic exploration involves the use of low frequency sound (< 100 Hz) to probe deep into the seabed. Despite the relatively poor resolution due to their long wavelength, low frequency sounds are preferred because high frequencies are heavily attenuated when they travel through the seabed. Sound sources used include airguns, vibroseis and explosives.

Weather and Climate Observation

Acoustic sensors can be used to monitor the sound made by wind and precipitation. For example, an acoustic rain gauge is described by Nystuen. Lightning strikes can also be detected. Acoustic thermometry of ocean climate (ATOC) uses low frequency sound to measure the global ocean temperature.

Oceanography

Large scale ocean features can be detected by acoustic tomography. Bottom characteristics can be measured by side-scan sonar and sub-bottom profiling.

Marine Biology

Due to its excellent propagation properties, underwater sound is used as a tool to aid the study of marine life, from microplankton to the blue whale. Echo sounders are often used to provide data on marine life abundance, distribution, and behavior information. Echo sounders, also referred to as hydroacoustics is also used for fish location, quantity, size, and biomass.

Acoustic telemetry is also used for monitoring fish and marine wildlife. An acoustic transmitter is attached to the fish (sometimes internally) while an array of receivers listen to the information conveyed by the sound wave. This enables the researchers to track the movements of individuals in a small-medium scale.

Pistol shrimp create sonoluminescent cavitation bubbles that reach up to 5,000 K (4,700 °C).

Particle Physics

A neutrino is a fundamental particle that interacts very weakly with other matter. For this reason, it requires detection apparatus on a very large scale, and the ocean is sometimes used for this purpose. In particular, it is thought that ultra-high energy neutrinos in seawater can be detected acoustically.

Psychoacoustics

Psychoacoustics is essentially the study of the perception of sound. This includes how we listen, our psychological responses, and the physiological impact of music and sound on the human nervous system.

In the realm of psychoacoustics, the terms music, sound, frequency, and vibration are interchangeable, because they are different approximations of the same essence. The study of psychoacoustics dissects the listening experience.

Traditionally, psychoacoustics is broadly defined as "pertaining to the perception of sound and the production of speech." The abundant research that has been done in the field has focused primarily on the exploration of speech and of the psychological effects of music therapy. Currently, however, there is renewed interest in sound as vibration.

An important distinction is the difference between a psychological and a neurological perception. A song or melody associated with childhood, a teenage romance, or some peak emotional experience creates a memory-based psychological reaction. There is also a physiological response to sounds, however. Slightly detuned tones can cause brain waves to speed up or slow down, for instance. Additionally, soundtracks that are filtered and gated (this is a sophisticated engineering process) create a random sonic event. It triggers an active listening response and thus tonifies the auditory mechanism, including the tiny muscles of the middle ear. As a result, sounds are perceived more accurately, and speech and communication skills improve. While a psychological response may occur with filtered and gated sounds, or detuned tones, the primary effect is physiological, or neurological, in nature.

Research on the neurological component of sound is currently attracting many to the field of psychoacoustics. A growing school of thought — based on the teachings of the Dr. Alfred Tomatis — values the examination of both neurological and psychological effects of resonance and frequencies on the human body.

The ear's primary purpose is to recycle sound and so recharge our inner batteries. According to Tomatis, the ear's first function in utero is to govern the growth of the rest of the physical organism. After birth, sound is to the nervous system what food is to our physical bodies: Food provides nourishment at the cellular level of the organism, and sound feeds us the electrical impulses that charge the neocortex. Indeed, psychoacoustics cannot be described at all without reference to the man known as the "Einstein of the ear."

In the realm of application-specific music and sound, psychoacoustically-designed soundtracks revolve around the following concepts and techniques:

- Intentionality (focused application for specific benefit).

- Resonance (tone).

- Entrainment (rhythm).

- Pattern Identification (active listening or passive hearing).

- Sonic Neurotechnologies (highly specialized sound processing).

Resonance and Entrainment

Consider the following: Anything that moves has a vibration. Though invisible, every aspect of our material world at the atomic level moves constantly. Wherever there is motion, there is frequency. Though inaudible at times, all frequencies make a sound. All sounds resonate and can affect one another. In the spectrum of sound — from the movement of atomic particles to the sensory phenomenon we call music — there is a chain of vibration:

- All atomic matter vibrates.

- Frequency is the speed at which matter vibrates.

- The frequency of vibration creates sound (sometimes inaudible).

- Sounds can be molded into music.

This chain explains the omnipresence of sound.

Resonance is the single most important concept in understanding the constructive or destructive role of sound in your life. Entrainment, sympathetic vibration, resonant frequencies, and resonant systems all fall under the rubric of resonance. Resonance can be broadly defined as "the impact of one vibration on another." Literally, it means "to send again, to echo." To resonate is to "re-sound." Something external sets something else into motion, or changes its vibratory rate. This can have many different effects — some subtle and some not so.

From iceburgs to airport construction to the human body, soundwaves have the capacity to alter, to actually shift frequency. Simply put, sound is a powerful — yet often ignored — medium for change.

Another fascinating and important aspect of resonance is the process of entrainment. Entrainment, in the context of psychoacoustics, concerns changing the rate of brain waves, breaths, or heartbeats from one speed to another through exposure to external, periodic rhythms.

The most common example of entrainment is tapping your feet to the external rhythm of music. Just try keeping your foot or your head still when you are around fun, up-tempo rhythms. You will see that it is almost an involuntary motor response. However, tapping your feet or bopping your head to external rhythms is just the tip of the iceberg. While your feet might be jitterbugging, your nervous system may be getting a terrible case of the jitters.

Rhythmic entrainment is contagious: If the brain doesn't resonate with a rhythm, neither will the breath or heart rate. In this context, rhythm takes on new meanings. Not only is it entertaining, but rhythmic entrainment is a potent sonic tool as well — be it for motor function or other autonomic processes such as brainwave, heart, and breath rates. Alter one pulse (such as brain waves) with music, and the other major pulses (heart and breath) will dutifully follow.

When it comes to the intentional applications of music, the entrainment effect completes the circle of the chain of vibration: Atomic matter —> vibration —> frequency —> sound —> sympathetic vibration (resonance) —> entrainment.

Music alters the performance of the nervous system primarily because of entrainment. Entrainment is the rhythmic manifestation of resonance. With entrainment, a stronger external pulse does not just activate another pulse but actually causes the latter to move out of its own resonant frequency to match it.

Understanding the interlocking concepts of resonance and entrainment enables us to grasp the way external tone and rhythm can heal or create havoc. Sound affects glass and concrete as well as brain waves, motor response, and organic cells.

Pattern Identification

Simply put, pattern identification is one of the brain's analytical processes. Identifying a pattern (visual, auditory, odiferous, kinesthetic) enables cerebral attention to shift from active awareness to passive acknowledgement. Listening and looking are active functions; hearing and seeing are passive.

In active listening mode, the middle ear function is highly engaged while the brain seeks to identify a pattern. Once an auditory pattern is found, passive hearing begins. Habituation sets in and the brain focuses on other things. There are specific times when active listening or passive hearing is preferable. Active listening stimulates the nervous system. Passive hearing is neutral or "discharging."

Sonic Neuro-technologies

Representing two distinct approaches to therapeutic sound, filtration/gating (F/G) and binaural beat frequencies (BBFs) currently define the growing field of "sonic neurotechnologies." This phrase was coined by Joshua Leeds to describe the arena of soundwork that depends on the precise mechanical manipulation of soundwaves to bring about desired changes in the psyche and physical body. Two diverse approaches to the processing of sound frequencies hold great interest and are used on some of the audio programs in Sound Remedies.

Filtration/gating (F/G) techniques have been honed in Tomatis clinics worldwide. By gradually gating and filtering out the lower range of music (sometimes up to 8000 Hz), and then adding the frequencies back in, a retraining of the auditory processing system occurs. The effects of filtration and gating are felt on a psychological, neurodevelopmental, and physical level. The application of sound stimulation has been effective in the remediation of many neurodevelopmental issues. Children and adults with learning/attention difficulties, developmental delays, auditory processing problems, sensory integration and perceptual challenges have experienced profound improvement.

Another approach to sound processing is the field of binaural beat frequencies (BBFs). By listening through stereo headphones to slightly detuned tones (i.e., sound frequencies that differ by a prescribed number of Hz), sonic brainwave entrainment takes place. Facilitating a specific range of brainwave states may assist in arenas such as pain reduction, enhanced creativity, or accelerated learning.

These two sonic neurotechnologies — used separately — have roots in neurology, physiology, and psychology. They must be used carefully and wisely. BBF and F/G soundtracks can be powerful tools. Consequently, proper consideration must always be afforded.

Sound products with BBFs or F/G contribute to health and wellness, but they are never intended to replace medical diagnosis or treatment. Do not drive or operate machinery while listening to sound programs that use these methedologies.

The therapeutic use of sound, like any new tool, requires discipline, education, and strict observance of ethical standards. There is currently no established licensure in the use of sonic neurotechnologies. Therefore the onus of responsibility for handling the changes that occur as a consequence of the application of these methods (most specifically, filtration/gating) falls on the practitioner. Sound is a marvelous adjunct to an existing profession. Therapists and educators will do well in performing due diligence and acquiring proper training.

Sound Stimulation with Filtration/Gating

In the broadest definition, sound stimulation can be defined as the excitement of the nervous system by auditory information. Sound stimulation auditory retraining narrows the focus. In this context, a precise application of electronically processed sound, through headphones, can have the effect of retraining the auditory mechanism to take in a wider spectrum of sound frequencies. An ear that cannot process tone properly is a problem of great magnitude. As discussed in previous chapters, sufficient auditory tonal processing is a prerequisite to normal auditory sequential processing.

- Auditory tonal processing (ATP) may be defined as the ability to differentiate between the tones utilized in language.

- Auditory sequential processing (ASP) is the ability to link pieces of auditory information together.

Auditory tonal processing is a basis for more complex levels of auditory sequential processing. ASP is the ability to receive, hold, process, and utilize auditory information using our short-term memory. As the foundation for short-term memory, ASP is one of the building blocks of thinking.

Sequential processing functions are fundamental to speech, language, learning, and other perceptual skills. The ability to interpret sound efficiently provides the neurological foundation for these sequential functions. Per neurodevelopmental specialist Robert J. Doman Jr., "many people who have experienced auditory processing deficits have seen their sequential functions return and improve when proper tonal processing is restored".

The primary sound application used in the remediation of impaired tonal processing was created by Alfred Tomatis. Further discussions cannot take place without absolute acknowledgment of his pioneering research. The current field of sound stimulation auditory retraining evolves from Tomatis's discoveries of the powerful effect of filtration and gating of sound.

References

- Till, rupert (2014). 'Sound archaeology: an interdisciplinary perspective', in archaeoacoustics: the archaeology of sound, linda einix (ed). Malta: ots foundation. Isbn 978-1497591264

- Aeroacoustics, multiphysics: comsol.co.in, Retrieved 28 April, 2019

- Till, Rupert (2011-01-20). "Songs of the Stones: An Investigation into the Acoustic History and Culture of Stonehenge doi:10.5429/2079-3871(2010)v1i2.10en". IASPM@Journal. 1 (2): 1–18. doi:10.5429/2079-3871(2010)v1i2.10en. ISSN 2079-3871

- Audrey. "the history of reverse reverb". Retrieved 5 january 2018

- Atti, Andreas Spanias, Ted Painter, Venkatraman (2006). Audio signal processing and coding ([Online-Ausg.] ed.). Hoboken, NJ: John Wiley & Sons. p. 464. ISBN 0-471-79147-4

- Psychoacoustics-defined: thepowerofsound.net, Retrieved 13 July, 2019

- D. Lohse, b. Schmitz & m. Versluis (2001). "snapping shrimp make flashing bubbles". Nature. 413 (6855): 477–478. Bibcode:2001natur.413..477l. Doi:10.1038/35097152. Pmid 11586346

3

Key Concepts in Acoustics

Some of the key concepts of acoustics are acoustic attenuation, acoustic impedance, acoustic absorption, soundproofing and mode conversion. This chapter closely examines these concepts of acoustics to provide a thorough understanding of the subject.

Attenuation of Sound Waves

When sound travels through a medium, its intensity diminishes with distance. In idealized materials, sound pressure (signal amplitude) is only reduced by the spreading of the wave. Natural materials, however, all produce an effect which further weakens the sound. This further weakening results from scattering and absorption. Scattering is the reflection of the sound in directions other than its original direction of propagation. Absorption is the conversion of the sound energy to other forms of energy. The combined effect of scattering and absorption is called attenuation. Ultrasonic attenuation is the decay rate of the wave as it propagates through material.

Attenuation of sound within a material itself is often not of intrinsic interest. However, natural properties and loading conditions can be related to attenuation. Attenuation often serves as a measurement tool that leads to the formation of theories to explain physical or chemical phenomenon that decreases the ultrasonic intensity.

The amplitude change of a decaying plane wave can be expressed as:

$$A \quad A e^{-\alpha z}$$

In this expression A_o is the unattenuated amplitude of the propagating wave at some location. The amplitude A is the reduced amplitude after the wave has traveled a distance z from that initial location. The quantity α is the attenuation coefficient of the wave traveling in the z-direction. The dimensions of α are nepers/length, where a neper is a dimensionless quantity. The term e is the exponential (or Napier's constant) which is equal to approximately 2.71828.

The units of the attenuation value in Nepers per meter (Np/m) can be converted to decibels/length by dividing by 0.1151. Decibels is a more common unit when relating the amplitudes of two signals.

Attenuation is generally proportional to the square of sound frequency. Quoted values of attenuation are often given for a single frequency, or an attenuation value averaged over many frequencies may be given. Also, the actual value of the attenuation coefficient for a given material is highly dependent on the way in which the material was manufactured. Thus, quoted values of attenuation only give a rough indication of the attenuation and should not be automatically trusted. Generally, a reliable value of attenuation can only be obtained by determining the attenuation experimentally for the particular material being used.

Attenuation can be determined by evaluating the multiple backwall reflections seen in a typical A-scan display. The number of decibels between two adjacent signals is measured and this value is divided by the time interval between them. This calculation produces a attenuation coefficient in decibels per unit time Ut. This value can be converted to nepers/length by the following equation:

$$\alpha = \frac{0.1151}{v} U_t$$

Where v is the velocity of sound in meters per second and Ut is in decibels per second.

Acoustic Impedance

Acoustic impedance and specific acoustic impedance are measures of the opposition that a system presents to the acoustic flow resulting from an acoustic pressure applied to the system. The SI unit of acoustic impedance is the pascal second per cubic metre (Pa·s/m³) or the rayl per square metre (rayl/m²), while that of specific acoustic impedance is the pascal second per metre (Pa·s/m) or the rayl. The symbol rayl denotes the MKS rayl. There is a close analogy with electrical impedance, which measures the opposition that a system presents to the electrical flow resulting from an electrical voltage applied to the system.

For a linear time-invariant system, the relationship between the acoustic pressure applied to the system and the resulting acoustic volume flow rate through a surface perpendicular to the direction of that pressure at its point of application is given by:

$$p(t) = [R * Q](t),$$

or equivalently by:

$$Q(t) = [G * p](t),$$

where:

- p is the acoustic pressure;
- Q is the acoustic volume flow rate;
- $*$ is the convolution operator;

- R is the acoustic resistance in the *time domain*;
- $G = R^{-1}$ is the acoustic conductance in the *time domain* (R^{-1} is the convolution inverse of R).

Acoustic impedance, denoted Z, is the Laplace transform, or the Fourier transform, or the analytic representation of *time domain* acoustic resistance:

$$Z(s) \overset{\text{def}}{=} \mathcal{L}[R](s) = \frac{\mathcal{L}[p](s)}{\mathcal{L}[Q](s)},$$

$$Z(\omega) \overset{\text{def}}{=} \mathcal{F}[R](\omega) = \frac{\mathcal{F}[p](\omega)}{\mathcal{F}[Q](\omega)},$$

$$Z(t) \overset{\text{def}}{=} R_a(t) = \frac{1}{2}\left[p_a * \left(Q^{-1}\right)_a \right](t),$$

where:

- $\mathcal{L}$ is the Laplace transform operator;
- $\mathcal{F}$ is the Fourier transform operator;
- Subscript "a" is the analytic representation operator;
- Q^{-1} is the convolution inverse of Q.

Acoustic resistance, denoted R, and acoustic reactance, denoted X, are the real part and imaginary part of acoustic impedance respectively:

$$Z(s) = R(s) + iX(s),$$

$$Z(\omega) = R(\omega) + iX(\omega),$$

$$Z(t) = R(t) + iX(t),$$

where:

- i is the imaginary unit;
- In $Z(s)$, $R(s)$ is *not* the Laplace transform of the time domain acoustic resistance $R(t)$, $Z(s)$ is;
- In $Z(\omega)$, $R(\omega)$ is *not* the Fourier transform of the time domain acoustic resistance $R(t)$, $Z(\omega)$ is;
- In $Z(t)$, $R(t)$ is the time domain acoustic resistance and $X(t)$ is the Hilbert transform of the time domain acoustic resistance $R(t)$, according to the definition of the analytic representation.

Inductive acoustic reactance, denoted X_L, and capacitive acoustic reactance, denoted X_C, are the positive part and negative part of acoustic reactance respectively:

$$X(s) = X_L(s) - X_C(s),$$

$$X(\omega) = X_L(\omega) - X_C(\omega),$$

$$X(t) = X_L(t) - X_C(t).$$

Acoustic admittance, denoted Y, is the Laplace transform, or the Fourier transform, or the analytic representation of *time domain* acoustic conductance:

$$Y(s) \stackrel{\text{def}}{=} \mathcal{L}[G](s) = \frac{1}{Z(s)} = \frac{\mathcal{L}[Q](s)}{\mathcal{L}[p](s)},$$

$$Y(w) \stackrel{\text{def}}{=} \mathcal{L}[G](w) = \frac{1}{Z(w)} = \frac{\mathcal{L}[Q](w)}{\mathcal{L}[p](w)},$$

$$Y(t) \stackrel{\text{def}}{=} G_a(t) = Z^{-1}(t) = \frac{1}{2}\left[Q_a * \left(p^{-1}\right)_a\right](t),$$

where:

- Z^{-1} is the convolution inverse of Z,

- p^{-1} is the convolution inverse of p.

Acoustic conductance, denoted G, and acoustic susceptance, denoted B, are the real part and imaginary part of acoustic admittance respectively:

$$Y(s) = G(s) + iB(s),$$

$$Y(\omega) = G(\omega) + iB(\omega),$$

$$Y(t) = G(t) + iB(t),$$

where:

- In $Y(s)$, $G(s)$ is *not* the Laplace transform of the time domain acoustic conductance $G(t)$, $Y(s)$ is,

- In $Y(\omega)$, $G(\omega)$ is *not* the Fourier transform of the time domain acoustic conductance $G(t)$, $Y(\omega)$ is,

- In $Y(t)$, $G(t)$ is the time domain acoustic conductance and $B(t)$ is the Hilbert transform of the time domain acoustic conductance $G(t)$, according to the definition of the analytic representation.

Acoustic resistance represents the energy transfer of an acoustic wave. The pressure and motion are in phase, so work is done on the medium ahead of the wave; as well, it represents the pressure that is out of phase with the motion and causes no average energy transfer. For example, a closed bulb connected to an organ pipe will have air moving into it and pressure, but they are out of phase so no net energy is transmitted into it. While the pressure rises, air moves in, and while it falls, it moves out, but the average pressure when the air moves in is the same as that when it moves out, so the power flows back and forth but with no time averaged energy transfer. A further electrical analogy is a capacitor connected across a power line: current flows through the capacitor but it is out of phase with the voltage, so no net power is transmitted into it.

Specific Acoustic Impedance

For a linear time-invariant system, the relationship between the acoustic pressure applied to the system and the resulting particle velocity in the direction of that pressure at its point of application is given by:

$$p(t) = [r * v](t),$$

or equivalently by:

$$v(t) = [g * p](t),$$

where:

- p is the acoustic pressure,
- v is the particle velocity,
- r is the specific acoustic resistance in the *time domain*,
- $g = r^{-1}$ is the specific acoustic conductance in the *time domain* (r^{-1} is the convolution inverse of r).

Specific acoustic impedance, denoted z is the Laplace transform, or the Fourier transform, or the analytic representation of *time domain* specific acoustic resistance:

$$z(s) \overset{\text{def}}{=} \mathcal{L}[r](s) = \frac{\mathcal{L}[p](s)}{\mathcal{L}[v](s)},$$

$$z(\omega) \overset{\text{def}}{=} \mathcal{F}[r](\omega) = \frac{\mathcal{F}[p](\omega)}{\mathcal{F}[v](\omega)},$$

$$z(t) \overset{\text{def}}{=} r_a(t) = \frac{1}{2}\left[p_a * \left(v^{-1}\right)_a \right](t),$$

where v^{-1} is the convolution inverse of v.

Specific acoustic resistance, denoted r, and specific acoustic reactance, denoted x, are the real part and imaginary part of specific acoustic impedance respectively:

$$z(s) = r(s) + ix(s),$$

$$z(\omega) = r(\omega) + ix(\omega),$$

$$z(t) = r(t) + ix(t),$$

where:

- In $z(s)$, $r(s)$ is *not* the Laplace transform of the time domain specific acoustic resistance $r(t)$, $z(s)$ is;
- In $z(\omega)$, $r(\omega)$ is *not* the Fourier transform of the time domain specific acoustic resistance $r(t)$, $z(\omega)$ is;

- In $z(t)$, $r(t)$ is the time domain specific acoustic resistance and $x(t)$ is the Hilbert transform of the time domain specific acoustic resistance $r(t)$, according to the definition of the analytic representation.

Specific inductive acoustic reactance, denoted x_L, and specific capacitive acoustic reactance, denoted x_C, are the positive part and negative part of specific acoustic reactance respectively:

$$x(s) = x_L(s) - x_C(s),$$

$$x(\omega) = x_L(\omega) - x_C(\omega),$$

$$x(t) = x_L(t) - x_C(t).$$

Specific acoustic admittance, denoted y, is the Laplace transform, or the Fourier transform, or the analytic representation of *time domain* specific acoustic conductance:

$$y(s) \overset{\text{def}}{=} \mathcal{L}[g](s) = \frac{1}{z(s)} = \frac{\mathcal{L}[v](s)}{\mathcal{L}[p](s)},$$

$$y(\omega) \overset{\text{def}}{=} \mathcal{F}[g](\omega) = \frac{1}{z(\omega)} = \frac{\mathcal{F}[v](\omega)}{\mathcal{F}[p](\omega)},$$

$$y(t) \overset{\text{def}}{=} g_a(t) = z^{-1}(t) = \frac{1}{2}\left[v_a * \left(p^{-1}\right)_a\right](t),$$

where:

- z^{-1} is the convolution inverse of z;

- p^{-1} is the convolution inverse of p.

Specific acoustic conductance, denoted g, and specific acoustic susceptance, denoted b, are the real part and imaginary part of specific acoustic admittance respectively:

$$y(s) = g(s) + ib(s),$$

$$y(\omega) = g(\omega) + ib(\omega),$$

$$y(t) = g(t) + ib(t),$$

where:

- In $y(s)$, $g(s)$ is *not* the Laplace transform of the time domain acoustic conductance $g(t)$, $y(s)$ is;

- In $y(\omega)$, $g(\omega)$ is *not* the Fourier transform of the time domain acoustic conductance $g(t)$, $y(\omega)$ is;

- In $y(t)$, $g(t)$ is the time domain acoustic conductance and $b(t)$ is the Hilbert transform of the time domain acoustic conductance $g(t)$, according to the definition of the analytic representation.

Specific acoustic impedance z is an intensive property of a particular *medium* (e.g., the z of air or water can be specified); on the other hand, acoustic impedance Z is an extensive property of a particular *medium and geometry* (e.g., the Z of a particular duct filled with air can be specified).

Relationship

For a *one dimensional* wave passing through an aperture with area A, the acoustic volume flow rate Q is the volume of medium passing per second through the aperture; if the acoustic flow moves a distance $dx = v\, dt$, then the volume of medium passing through is $dV = A\, dx$, so:

$$Q = \frac{dV}{dt} = A\frac{dx}{dt} = Av.$$

Provided that the wave is only one-dimensional, it yields:

$$Z(s) = \frac{\mathcal{L}[p](s)}{\mathcal{L}[Q](s)} = \frac{\mathcal{L}[p](s)}{A\mathcal{L}[v](s)} = \frac{z(s)}{A},$$

$$Z(\omega) = \frac{\mathcal{F}[p](\omega)}{\mathcal{F}[Q](\omega)} = \frac{\mathcal{F}[p](\omega)}{A\mathcal{F}[v](\omega)} = \frac{z(\omega)}{A},$$

$$Z(t) = \frac{1}{2}\left[p_a * \left(Q^{-1}\right)_a \right](t) = \frac{1}{2}\left[p_a * \left(\frac{v^{-1}}{A}\right)_a \right](t) = \frac{z(t)}{A}.$$

Characteristic Specific Acoustic Impedance

The constitutive law of nondispersive linear acoustics in one dimension gives a relation between stress and strain:

$$p = -\rho c^2 \frac{\partial \delta}{\partial x},$$

where

- p is the acoustic pressure in the medium;
- ρ is the volumetric mass density of the medium;
- c is the speed of the sound waves traveling in the medium;
- δ is the particle displacement;
- x is the space variable along the direction of propagation of the sound waves.

This equation is valid both for fluids and solids. In

- Fluids, $\rho c^2 = K$ (K stands for the bulk modulus);
- solids, $\rho c^2 = K + 4/3\, G$ (G stands for the shear modulus) for longitudinal waves and $\rho c^2 = G$ for transverse waves.

Newton's second law applied locally in the medium gives:

$$\rho \frac{\partial^2 \delta}{\partial t^2} = -\frac{\partial p}{\partial x}.$$

Combining this equation with the previous one yields the one-dimensional wave equation:

$$\frac{\partial^2 \delta}{\partial t^2} = c^2 \frac{\partial^2 \delta}{\partial x^2}.$$

The plane waves:

$$\delta(\mathbf{r}, t) = \delta(x, t)$$

that are solutions of this wave equation are composed of the sum of *two progressive plane waves* traveling along *x* with the same speed and *in opposite ways*:

$$\delta(\mathbf{r}, t) = f(x - ct) + g(x + ct)$$

from which can be derived:

$$v(\mathbf{r}, t) = \frac{\partial \delta}{\partial t}(\mathbf{r}, t) = -c[f'(x - ct) - g'(x + ct)],$$

$$p(\mathbf{r}, t) = -\rho c^2 \frac{\partial \delta}{\partial x}(\mathbf{r}, t) = -\rho c^2 \left[f'(x - ct) + g'(x + ct) \right].$$

For *progressive* plane waves:

$$\begin{cases} p(\mathbf{r}, t) = -\rho c^2 f'(x - ct) \\ v(\mathbf{r}, t) = -c f'(x - ct) \end{cases}$$

or,

$$\begin{cases} p(\mathbf{r}, t) = -\rho c^2 g'(x + ct) \\ v(\mathbf{r}, t) = c g'(x + ct). \end{cases}$$

Finally, the specific acoustic impedance *z* is:

$$z(\mathbf{r}, s) = \frac{\mathcal{L}[p](\mathbf{r}, s)}{\mathcal{L}[v](\mathbf{r}, s)} = \pm \rho c,$$

$$z(\mathbf{r}, \omega) = \frac{\mathcal{F}[p](\mathbf{r}, \omega)}{\mathcal{F}[v](\mathbf{r}, \omega)} = \pm \rho c,$$

$$z(\mathbf{r}, t) = \frac{1}{2} \left[p_a * \left(v^{-1} \right)_a \right] (\mathbf{r}, t) = \pm \rho c.$$

The absolute value of this specific acoustic impedance is often called characteristic specific acoustic impedance and denoted z_0:

$$z_0 = \rho c.$$

The equations also show that:

$$\frac{p(\mathbf{r}, t)}{v(\mathbf{r}, t)} = \pm \rho c = \pm z_0.$$

z_0 varies greatly among media, especially between gas and condensed phases; for instance, water is 800 times denser than air and its speed of sound is 4.3 times as fast as that of air, and so the specific acoustic impedance of water is 3,500 times higher than that of air. This means that a sound in water with a given pressure amplitude is 3,500 times less intense than one in air with the same pressure—because air, with its lower z_0, moves with a much greater velocity and displacement amplitude than water; reciprocally, if a sound in water and another in air have the same intensity, then the pressure is much smaller in air. These variations lead to important differences between room acoustics or atmospheric acoustics on the one hand, and underwater acoustics on the other.

Effect of Temperature

Temperature acts on speed of sound and mass density and thus on specific acoustic impedance.

Effect of temperature on properties of air			
Temperature T (°C)	Speed of sound c (m/s)	Density of air ρ (kg/m³)	Characteristic specific acoustic impedance z_0 (Pa·s/m)
35	351.88	1.1455	403.2
30	349.02	1.1644	406.5
25	346.13	1.1839	409.4
20	343.21	1.2041	413.3
15	340.27	1.2250	416.9
10	337.31	1.2466	420.5
5	334.32	1.2690	424.3
0	331.30	1.2922	428.0
−5	328.25	1.3163	432.1
−10	325.18	1.3413	436.1
−15	322.07	1.3673	440.3
−20	318.94	1.3943	444.6
−25	315.77	1.4224	449.1

Characteristic Acoustic Impedance

For a *one dimensional* wave passing through an aperture with area A, $Z = z/A$, so if the wave is a progressive plane wave, then:

$$Z(\mathbf{r}, s) = \pm \frac{\rho c}{A},$$

$$Z(\mathbf{r}, \omega) = \pm \frac{\rho c}{A},$$

$$Z(\mathbf{r}, t) = \pm \frac{\rho c}{A}.$$

The absolute value of this acoustic impedance is often called characteristic acoustic impedance and denoted Z_0:

$$Z_0 = \frac{\rho c}{A}.$$

and the characteristic specific acoustic impedance is:

$$\frac{p(\mathbf{r}, t)}{Q(\mathbf{r}, t)} = \pm \frac{\rho c}{A} = \pm Z_0.$$

If the aperture with area A is the start of a pipe and a plane wave is sent into the pipe, the wave passing through the aperture is a progressive plane wave in the absence of reflections, and the usually reflections from the other end of the pipe, whether open or closed, are the sum of waves travelling from one end to the other. (It is possible to have no reflections when the pipe is very long, because of the long time taken for the reflected waves to return, and their attenuation through losses at the pipe wall). Such reflections and resultant standing waves are very important in the design and operation of musical wind instruments.

Acoustic Attenuation

Acoustic attenuation is a measure of the energy loss of sound propagation in media. Most media have viscosity, and are therefore not ideal media. When sound propagates in such media, there is always thermal consumption of energy caused by viscosity. For inhomogeneous media, besides media viscosity, acoustic scattering is another main reason for removal of acoustic energy. Acoustic attenuation in a lossy medium plays an important role in many scientific researches and engineering fields, such as medical ultrasonography, vibration and noise reduction.

Power-law Frequency-dependent Acoustic Attenuation

Many experimental and field measurements show that the acoustic attenuation coefficient of a wide range of viscoelastic materials, such as soft tissue, polymers, soil and porous rock, can be expressed as the following power law with respect to frequency:

$$P(x + \Delta x) = P(x)e^{-\alpha(\omega)\Delta x}, \alpha(\omega) = \alpha_0 \omega^\eta$$

Where ω is the angular frequency, P the pressure, Δx the wave propagation distance, $\alpha(\omega)$ the attenuation coefficient, α_0 and frequency dependent exponent η are real non-negative material parameters obtained by fitting experimental data and the value of η ranges from 0 to 2. Acoustic attenuation in water, many metals and crystalline materials are frequency-squared dependent, namely $\eta = 2$. In contrast, it is widely noted that the frequency dependent exponent η of viscoelastic materials is between 0 and 2. For example, the exponent η of sediment, soil and rock is about 1, and the exponent η of most soft tissues is between 1 and 2.

The classical dissipative acoustic wave propagation equations are confined to the frequency-independent and frequency-squared dependent attenuation, such as damped wave equation and approximate thermoviscous wave equation. In recent decades, increasing attention and efforts are focused on developing accurate models to describe general power law frequency-dependent acoustic attenuation. Most of these recent frequency-dependent models are established via the analysis of the complex wave number and are then extended to transient wave propagation. The multiple relaxation model considers the power law viscosity underlying different molecular relaxation processes. Szabo proposed a time convolution integral dissipative acoustic wave equation. On the other hand, acoustic wave equations based on fractional derivative viscoelastic models are applied to describe the power law frequency dependent acoustic attenuation. Chen and Holm proposed the positive fractional derivative modified Szabo's wave equation and the fractional Laplacian wave equation.

The phenomenon of attenuation obeying a frequency power-law may be described using a causal wave equation, derived from a fractional constitutive equation between stress and strain. This wave equation incorporates fractional time derivatives:

$$\nabla^2 u - \frac{1}{c_0^2}\frac{\partial^2 u}{\partial t^2} + \tau_\sigma^\alpha \frac{\partial^\alpha}{\partial t^\alpha}\nabla^2 u - \frac{\tau_\epsilon^\beta}{c_0^2}\frac{\partial^{\beta+2} u}{\partial t^{\beta+2}} = 0.$$

Such fractional derivative models are linked to the commonly recognized hypothesis that multiple relaxation phenomena give rise to the attenuation measured in complex media.

For frequency band-limited waves, Ref. describes a model-based method to attain causal power-law attenuation using a set of discrete relaxation mechanisms within the Nachman et al. framework.

Acoustic Absorption

Acoustic absorption refers to the process by which a material, structure, or object takes in sound energy when sound waves are encountered, as opposed to reflecting the energy. Part of the absorbed energy is transformed into heat and part is transmitted through the absorbing body. The energy transformed into heat is said to have been 'lost'.

When sound from a loudspeaker collides with the walls of a room part of the sound's energy is reflected, part is transmitted, and part is absorbed into the walls. Just as the acoustic energy was transmitted through the air as pressure differentials (or deformations), the acoustic energy travels through the material which makes up the wall in the same manner. Deformation causes mechanical losses via conversion of part of the sound energy into heat, resulting in acoustic attenuation, mostly due to the wall's viscosity. Similar attenuation mechanisms apply for the air and any other medium through which sound travels.

The fraction of sound absorbed is governed by the acoustic impedances of both media and is a function of frequency and the incident angle. Size and shape can influence the sound wave's behavior if they interact with its wavelength, giving rise to wave phenomena such as standing waves and diffraction.

Acoustic absorption is of particular interest in soundproofing. Soundproofing aims to absorb as much sound energy (often in particular frequencies) as possible converting it into heat or transmitting it away from a certain location.

In general, soft, pliable, or porous materials (like cloths) serve as good acoustic insulators - absorbing most sound, whereas dense, hard, impenetrable materials (such as metals) reflect most.

How well a room absorbs sound is quantified by the effective absorption area of the walls, also named total absorption area. This is calculated using its dimensions and the absorption coefficients of the walls. The total absorption is expressed in Sabins and is useful in, for instance, determining the reverberation time of auditoria. Absorption coefficients can be measured using a reverberation room, which is the opposite of an anechoic chamber.

Absorption coefficients of common materials					
Materials	Frequency (Hz)				
	125	250	500	1,000	2,000
Acoustic tile (ceiling)	.80	.90	.90	.95	.90
Brick	.03	.03	.03	.04	.05
Carpet over concrete	.08	.25	.60	.70	.72
Heavy curtains	.15	.35	.55	.75	.70
Marble	.01	.01	.01	.01	.02
Painted concrete	.10	.05	.06	.07	.09
Plaster on concrete	.10	.10	.08	.05	.05
Plywood on studs	.30	.20	.15	.10	.09
Smooth concrete	.01	.01	.01	.02	.02
Wood floor	.15	.11	.10	.07	.06

Applications

Acoustic absorption is critical in areas such as:

- Soundproofing
- Sound recording and reproduction
- Loudspeaker design
- Acoustic transmission lines

- Room acoustics
- Architectural acoustics
- Sonar
- Noise Barrier Walls

An anechoic chamber

Anechoic Chamber

An acoustic anechoic chamber is a room designed to absorb as much sound as possible. The walls consist of a number of baffles with highly absorptive material arranged in such a way that the fraction of sound they do reflect is directed towards another baffle instead of back into the room. This makes the chamber almost devoid of echos which is useful for measuring the sound pressure level of a source and for various other experiments and measurements. Anechoic chambers are expensive for several reasons and are therefore not common.

They must be isolated from outside influences (e.g., planes, trains, automobiles, snowmobiles, elevators, pumps, indeed any source of sound which may interfere with measurements inside the chamber) and they must be physically large. The first, environmental isolation, requires in most cases specially constructed, nearly always massive, and likewise thick, walls, floors, and ceilings. Such chambers are often built as spring supported isolated rooms within a larger building. The National Research Council in Canada has a modern anechoic chamber, and has posted a video on the Web, noting these as well as other constructional details. Doors must be specially made, sealing for them must be acoustically complete (no leaks around the edges), ventilation (if any) carefully managed, and lighting chosen to be silent.

The second requirement follows in part from the first and from the necessity of preventing reverberation inside the room from, say, a sound source being tested. Preventing echoes is almost always done with absorptive foam wedges on walls, floors and ceilings, and if they are to be effective at low frequencies, these must be physically large; the lower the frequencies to be absorbed, the larger they must be.

An anechoic chamber must therefore be large to accommodate those absorbers and isolation schemes, but still allow for space for experimental apparatus and units under test.

Electrical and Mechanical Analogy

The energy dissipated within a medium as sound travels through it is analogous to the energy dissipated in electrical resistors or that dissipated in mechanical dampers for mechanical motion transmission systems. All three are equivalent to the resistive part of a system of resistive and reactive elements. The resistive elements dissipate energy (irreversibly into heat) and the reactive elements store and release energy (reversibly, neglecting small losses). The reactive parts of an acoustic medium are determined by its bulk modulus and its density, analogous to respectively an electrical capacitor and an electrical inductor, and analogous to, respectively, a mechanical spring attached to a mass.

Note that since dissipation solely relies on the resistive element it is independent of frequency. In practice however the resistive element varies with frequency. For instance, vibrations of most materials change their physical structure and so their physical properties; the result is a change in the 'resistance' equivalence. Additionally, the cycle of compression and rarefaction exhibits hysteresis of pressure waves in most materials which is a function of frequency, so for every compression there is a rarefaction, and the total amount of energy dissipated due to hysteresis changes with frequency. Furthermore, some materials behave in a non-Newtonian way, which causes their viscosity to change with the rate of shear strain experienced during compression and rarefaction; again, this varies with frequency. Gasses and liquids generally exhibit less hysteresis than solid materials (e.g., sound waves cause adiabatic compression and rarefaction) and behave in a, mostly, Newtonian way.

Combined, the resistive and reactive properties of an acoustic medium form the acoustic impedance. The behaviour of sound waves encountering a different medium is dictated by the differing acoustic impedances. As with electrical impedances, there are matches and mismatches and energy will be transferred for certain frequencies (up to nearly 100%) whereas for others it could be mostly reflected (again, up to very large percentages).

In amplifier and loudspeaker design electrical impedances, mechanical impedances, and acoustic impedances of the system have to be balanced such that the frequency and phase response least alter the reproduced sound across a very broad spectrum whilst still producing adequate sound levels for the listener. Modelling acoustical systems using the same (or similar) techniques long used in electrical circuits gave acoustical designers a new and powerful design tool.

Soundproofing

Soundproofing is any means of reducing the sound pressure with respect to a specified sound source and receptor. There are several basic approaches to reducing sound: increasing the distance between source and receiver, using noise barriers to reflect or absorb the energy of the sound waves, using damping structures such as sound baffles, or using active antinoise sound generators.

Two distinct soundproofing problems may need to be considered when designing acoustic treatments—to improve the sound within a room, and reduce sound leakage to/from adjacent rooms or outdoors. Acoustic quieting and noise control can be used to limit unwanted noise. Soundproofing can suppress unwanted indirect sound waves such as reflections that cause echoes and resonances that cause reverberation. Soundproofing can reduce the transmission of unwanted direct sound waves from the source to an involuntary listener through the use of distance and intervening objects in the sound path.

Distance

The energy density of sound waves decreases as they spread out, so that increasing the distance between the receiver and source results in a progressively lesser intensity of sound at the receiver. In a normal three-dimensional setting, with a point source and point receptor, the intensity of sound waves will be attenuated according to the inverse square of the distance from the source.

Damping

Damping means to reduce resonance in the room, by absorption or redirection (reflection or diffusion). Absorption will reduce the overall sound level, whereas redirection makes unwanted sound harmless or even beneficial by reducing coherence. Damping can reduce the acoustic resonance in the air, or mechanical resonance in the structure of the room itself or things in the room.

Absorption

Absorbing sound spontaneously converts part of the sound energy to a very small amount of heat in the intervening object (the absorbing material), rather than sound being transmitted or reflected. There are several ways in which a material can absorb sound. The choice of sound absorbing material will be determined by the frequency distribution of noise to be absorbed and the acoustic absorption profile required.

Porous Absorbers

Porous absorbers, typically open cell rubber foams or melamine sponges, absorb noise by friction within the cell structure. Porous open cell foams are highly effective noise absorbers across a broad range of medium-high frequencies. Performance can be less impressive at lower frequencies.

The exact absorption profile of a porous open cell foam will be determined by a number of factors including the following:

- Cell size
- Material thickness
- Tortuosity
- Material density
- Porosity

Resonant Absorbers

Resonant panels, Helmholtz resonators and other resonant absorbers work by damping a sound wave as they reflect it. Unlike porous absorbers, resonant absorbers are most effective at low-medium frequencies and the absorption of resonant absorbers is always matched to a narrow frequency range.

Reflection

When sound waves hit a medium, the reflection of that sound is dependent on dissimilarity of the surfaces it comes in contact with. Sound hitting a concrete surface will result in a much different reflection than if sound were to hit a softer medium such as fiberglass. In an outdoor environment such as highway engineering, embankments or panelling are often used to reflect sound upwards into the sky.

Diffusion

If a specular reflection from a hard flat surface is giving a problematic echo then an acoustic diffuser may be applied to the surface. It will scatter sound in all directions. This is effective to eliminate pockets of noise in a room.

Room within a Room

A room within a room (RWAR) is one method of isolating sound and preventing it from transmitting to the outside world where it may be undesirable.

Most vibration/sound transfer from a room to the outside occurs through mechanical means. The vibration passes directly through the brick, woodwork and other solid structural elements. When it meets with an element such as a wall, ceiling, floor or window, which acts as a sounding board, the vibration is amplified and heard in the second space. A mechanical transmission is much faster, more efficient and may be more readily amplified than an airborne transmission of the same initial strength.

The use of acoustic foam and other absorbent means is less effective against this transmitted vibration. The user is advised to break the connection between the room that contains the noise source and the outside world. This is called acoustic decoupling. Ideal decoupling involves eliminating vibration transfer in both solid materials and in the air, so air-flow into the room is often controlled. This has safety implications: inside decoupled space, proper ventilation must be assured, and gas heaters cannot be used.

Noise Cancellation

Noise cancellation generators for active noise control are a relatively modern innovation. A microphone is used to pick up the sound that is then analyzed by a computer; then, sound waves with opposite polarity (180° phase at all frequencies) are output through a speaker, causing destructive interference and cancelling much of the noise.

Residential Soundproofing

Residential soundproofing aims to decrease or eliminate the effects of exterior noise. The main focus of residential soundproofing in existing structures is the windows and doors. Solid wood doors are a better sound barrier than hollow doors. Curtains can be used to dampen sound, either through use of heavy materials, or through the use of air chambers known as honeycombs. Single-, double- and triple-honeycomb designs achieve relatively greater degrees of sound damping. The primary soundproofing limit of curtains is the lack of a seal at the edge of the curtain, although this may be alleviated with the use of sealing features, such as hook and loop fastener, adhesive, magnets, or other materials. Thickness of glass will play a role when diagnosing sound leakage. Double-pane windows achieve somewhat greater sound damping than single-pane windows when well sealed into the opening of the window frame and wall.

Significant noise reduction can also be achieved by installing a second interior window. In this case the exterior window remains in place while a slider or hung window is installed within the same wall openings.

In the USA the FAA offers soundproofing for homes that fall within a noise contour where the average decibel level is 65 decibels. It is part of their Residential Sound Insulation Program. The program provides Solid-core wood entry doors plus windows and storm doors.

Commercial Soundproofing

Restaurants, schools, office businesses, and health care facilities use architectural acoustics to reduce noise for their customers. Commercial businesses sometimes use soundproofing technology, especially when they are an open office design. There are many reasons why a business might implement soundproofing for their office. One of the biggest hindrances in worker productivity are the distracting noises that comes from people talking such as on the phone, or with their co-workers and boss. Noise soundproofing is important in mitigating people from losing their concentration and focus from their work project. It is also important to keep confidential conversations secure to the intended listeners.

When trying to find places to install soundproofing, acoustic panels should be installed in office areas where lots of traffic corridors, circulation pathways, and open work areas are connected. Successful acoustic panel installations rely on three strategies and techniques to absorb sound, block sound transmission from one place to another, and cover and masking of the sound, positioned to avoid other services or block light.

Automotive Soundproofing

Automotive soundproofing aims to decrease or eliminate the effects of exterior noise, primarily engine, exhaust and tire noise across a wide frequency range. When constructing a vehicle which includes soundproofing, a panel damping material is fitted which reduces the vibration of the vehicle's body panels when they are excited by one of the many high energy sound sources caused when the vehicle is in use. There are many complex noises created within vehicles which change with the driving environment and speed at which the vehicle travels. Significant noise reductions of up to 8 dB can be achieved by installing a combination of different types of materials.

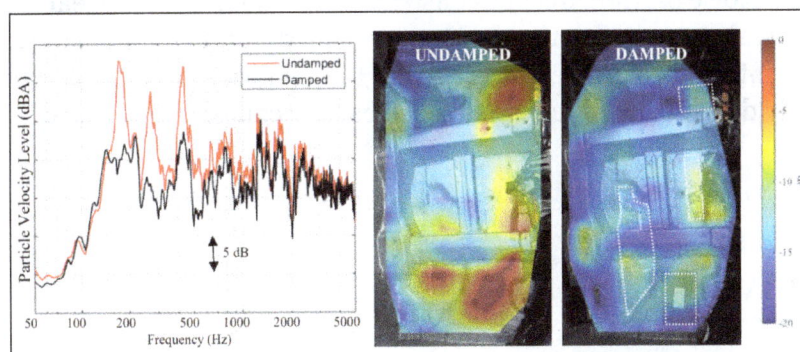

Spatially averaged particle velocity spectra (left) and broadband colormaps of a car floor without (middle) and with (right) a damping treatment.

The automotive environment limits the thickness of materials that can be used, but combinations of dampers, barriers, and absorbers are common. Common materials include felt, foam, polyester, and Polypropylene blend materials. Waterproofing may be necessary based on materials used. Acoustic foam can be applied in different areas of a vehicle during manufacture to reduce cabin noise. Foams also have cost and performance advantages in installation since foam material can expand and fill cavities after application and also prevent leaks and some gases from entering the vehicle. Vehicle soundproofing can reduce wind, engine, road, and tire noise. Vehicle soundproofing can reduce sound inside a vehicle from five to 20 decibels.

Surface damping materials are very effective at reducing structure borne noise. Passive damping materials have been used since the early 1960s in the aerospace industry. Over the years, advances in material manufacturing and the development of more efficient analytical and experimental tools to characterise complex dynamic behaviours enabled to expand the usage of these materials to the automotive industry. Nowadays, multiple viscoelastic damping pads are usually attached to the body in order to attenuate higher order structural panel modes that significantly contribute to the overall noise level inside the cabin. Traditionally, experimental techniques are used to optimise the size and location of damping treatments. In particular, laser vibrometer type tests are often conducted on body in white structures enabling the fast acquisition of a large number of measurement points with a good spatial resolution. However, testing a complete vehicle is mostly unfeasible, requiring to evaluate every subsystem individually, hence limiting the usability of this technology in a fast and efficient way. Alternatively, structural vibrations can also be acoustically measured using particle velocity sensors located near a vibrating structure. Several studies have revealed the potential of particle velocity sensors for characterising structural vibrations, which remarkably accelerates the entire testing process when combined with scanning techniques.

Noise Barriers as Exterior Soundproofing

Since the early 1970s, it has become common practice in the United States and other industrialized countries to engineer noise barriers along major highways to protect adjacent residents from intruding roadway noise. The Federal Highway Administration (FHWA) in conjunction with State Highway Administration (SHA) adopted Federal Regulation (23 CFR 772) requiring each state to adopt their own policy in regards to abatement of highway traffic noise. Engineering techniques have been developed to predict an effective geometry for the noise barrier design in a particular real world situation. Noise barriers may be constructed of wood, masonry, earth or a combination thereof. One of the earliest noise barrier designs was in Arlington, Virginia adjacent to Interstate 66, stemming from interests expressed by the Arlington Coalition on Transportation. Possibly the earliest scientifically designed and published noise barrier construction was in Los Altos, California in 1970.

Mode Conversion

When sound travels in a solid material, one form of wave energy can be transformed into another form. For example, when a longitudinal waves hits an interface at an angle, some of the energy can cause particle movement in the transverse direction to start a shear (transverse) wave. Mode conversion occurs when a wave encounters an interface between materials of different acoustic impedances and the incident angle is not normal to the interface. From the ray tracing movie below, it can be seen that since mode conversion occurs every time a wave encounters an interface at an angle, ultrasonic signals can become confusing at times.

When sound waves pass through an interface between materials having different acoustic veloci-ties, refraction takes place at the interface. The larger the difference in acoustic velocities between the two materials, the more the sound is refracted. Notice that the shear wave is not refracted as much as the longitudinal wave. This occurs because shear waves travel slower than longitudinal waves. Therefore, the velocity difference between the incident longitudinal wave and the shear wave is not as great as it is between the incident and refracted longitudinal waves. Also note that when a longitudinal wave is reflected inside the material, the reflected shear wave is reflected at a smaller angle than the reflected longitudinal wave. This is also due to the fact that the shear veloc-ity is less than the longitudinal velocity within a given material.

Snell's Law holds true for shear waves as well as longitudinal waves and can be written as follows.

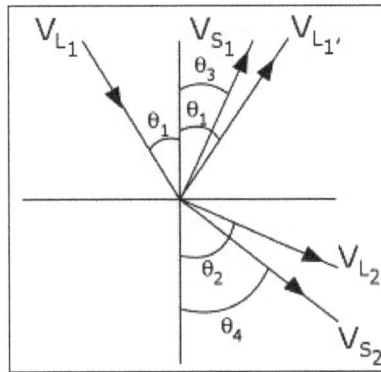

$$\frac{\sin\theta_1}{V_{L_1}} = \frac{\sin\theta_2}{V_{L_2}} = \frac{\sin}{V_{S_1}} = \frac{\sin\theta_4}{V_{S_2}}$$

Where:

- V_{L1} is the longitudinal wave velocity in material 1.

- V_{L2} is the longitudinal wave velocity in material 2.

- V_{S1} is the shear wave velocity in material 1.

- V_{S2} is the shear wave velocity in material 2.

The shear (transverse) wave ray path has been added. The ray paths of the waves can be ad-justed by clicking and dragging in the vicinity of the arrows. Values for the angles or the wave velocities can also be entered into the dialog boxes. It can be seen from the applet that when a wave moves from a slower to a faster material, there is an incident angle which makes the angle of refraction for the longitudinal wave 90 degrees. This surface following wave is some-time referred to as a creep wave and it is not very useful in NDT because it dampens out very rapidly.

Beyond the first critical angle, only the shear wave propagates into the material. For this reason, most angle beam transducers use a shear wave so that the signal is not complicated by having two waves present. In many cases there is also an incident angle that makes the angle of refraction for the shear wave 90 degrees. This is known as the second critical angle and at this point, all of the wave energy is reflected or refracted into a surface following shear wave or shear creep wave. Slightly beyond the second critical angle, surface waves will be generated.

References

- Attenuation, Physics, Ultrasonics, communitycollege, educationresources: nde-ed.org, Retrieved 15 January, 2019

- Kinsler, Lawrence; Frey, Austin; Coppens, Alan; Sanders, James (2000). Fundamentals of Acoustics. New York: John Wiley & Sons, Inc. ISBN 0-471-84789-5

- Chen, Yangkang; Ma, Jitao (May–June 2014). "Random noise attenuation by f-x empirical-mode decomposition predictive filtering". Geophysics. 79 (3): V81-V91. Bibcode:2014Geop...79...81C. Doi:10.1190/GEO2013-0080.1

- Good vibrations: the physics of music. Johns Hopkins University Press. P. 248. ISBN 9780801897078. Retrieved 4 January 2019

- "Reflection, Refraction, and Diffraction". Www.physicsclassroom.com. Retrieved 2017-07-10

- Wisniewski, Mary. "City wants more Midway-area homeowners to sign up for soundproofing". Chicagotribune.com. Retrieved 2017-02-05

- Modeconversion, Physics, Ultrasonics, communitycollege, educationresources: nde-ed.org, Retrieved 28 April, 2019

4
Characteristics of Sound

The vibration which typically propagates as an audible wave of pressure through a transmission medium is known as sound. Some of its common characteristics are amplitude, wavelength, timbre, frequency, energy, power and intensity. All these characteristics of sound have been carefully analyzed in this chapter.

Amplitude

Amplitude is the maximum displacement or distance moved by a point on a vibrating body or wave measured from its equilibrium position. It is equal to one-half the length of the vibration path. The amplitude of a pendulum is thus one-half the distance that the bob traverses in moving from one side to the other. Waves are generated by vibrating sources, their amplitude being proportional to the amplitude of the source.

For a transverse wave, such as the wave on a plucked string, amplitude is measured by the maximum displacement of any point on the string from its position when the string is at rest. For a longitudinal wave, such as a sound wave, amplitude is measured by the maximum displacement of a particle from its position of equilibrium. When the amplitude of a wave steadily decreases because its energy is being lost, it is said to be damped.

Wavelength

Wavelength is one of the more straightforward acoustics concepts to imagine. It is simply the size of a wave, measured from one peak to the next. If one imagines a sound wave as something like a water wave, then the wavelength is simply the distance from the crest of one wave to the next nearest crest. Thus, if the distance between two peaks is 1 m, then the wavelength is 1 m. There is a direct relation between wavelength, frequency, and sound speed. Namely, if we know the frequency (which is the number of wave repetitions per second, often given in Hertz, or Hz) and the sound speed (which is the speed the wave travels in meters per sec), then we can find the wavelength using the equation wavelength=speed/frequency.

Put another way, wavelength is the distance that a wave travels before the next wave starts. That means that at a given sound speed, as frequency gets higher, the time between repetitions decreases and the wavelength gets shorter, and vice versa.

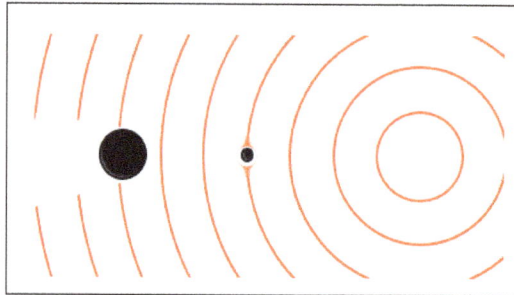

Waves can bend around small objects with ease, while larger objects may block those waves.

Wavelength is the essential quantity to know when trying to understand how waves move through the world. Long wavelengths bend around objects that are smaller then themselves, while short wavelengths reflect off of or are absorbed by those same objects. Thus, a sound with a wavelength of 3.4 cm in air (1,000 Hz) will not be hampered by an object that is less than 3.4 cm in diameter, but a larger object may interfere with or entirely block that wave.

Often people talk about "long" and "short," but what is really meant by these terms?" How does one draw the line between those admittedly fuzzy and highly subjective categories? To answer this question, we must understand the concept of scale. Scale is important throughout science, from biology to physics, though not all disciplines give it formal treatment.

Scale is relative. A pebble that may seem huge compared to an ant would be tiny next to an elephant.

To see how a physicist thinks of scale, think about this question: "How big is a 1-inch (2.54 cm) rock?" If your answer is "not very big," then you are thinking like a human. But imagine for a second that you were an ant. That rock would seem like a hill! If you were a mouse, the rock would be like a small boulder. If you are a human, the rock is merely a pebble. And if you were an elephant, the rock is like a tiny piece of gravel. Scale applies to more than just physical size, though. For example, you might have heard of a "geologic time scale," which refers to periods of time that are long enough to observe significant changes in the Earth itself. Almost every quantity can cover a wide scale, be that distance, pressure, time, or even money.

When talking about differences of scale, the term used most often is "order of magnitude," which is another way of saying power of ten. If one were to say something was "an order of magnitude larger," what they are really saying is that it is ten times larger than whatever it is being compared to. If something is two orders of magnitude larger, that would be 100 times the size. The order of magnitude is so important that it is part of scientific notation. For example, 5,000,000,000 meters (m) might be written as 5.0×10^9 m, with the 9 in the exponent being the order of magnitude.

While there are certainly exceptions, two quantities are considered comparable when they are within one order of magnitude of each other. Beyond that, we describe one as being "much larger" or "much smaller" than the other. While this may seem like just a semantic difference, in many physics equations having one quantity much smaller or much larger causes the math to clean up to much simpler forms, which corresponds to much simpler physical behavior.

While scale is important throughout science, there are few places where it is more apparent than with wavelength and sound. Wavelength of audible sounds, as it turns out, covers a very large range of scales. On the large end, you have low frequency waves with wavelengths of up to 17 meters (20 Hz), while the highest frequencies can be as small as 1.7 centimeters (20,000 Hz). Compare this to the wavelengths of visible light (430-790 nanometers), and not only do you find that sound covers a much wider range of scales (four orders of magnitude), but it also covers a range that is squarely at the scale of human experience.

For an example of how wavelength determines the behavior or sound, consider living in an apartment with a noisy next door neighbor. If that neighbor turns up their stereo, you might hear the bass clearly through the wall. The bass notes are low frequency, with very long wavelengths. Wavelengths so long, in fact, that the plaster drywall separating you from the music would be considered "very thin." The higher notes, on the other hand, get handily blocked by the wall, resulting in the "muffled" sound that is often associated with blocked sources.

Wavelength also determines how easy it is to find the direction of a sound. You may have heard, for example, that the placement of a subwoofer in a room does not make a big difference for a sound system. As it is often described, "low frequencies" are not directional. What this really means is that the wavelengths of low frequencies (which can be more than a meter long) are so large that the listener's head is much smaller than a wavelength. As a result, there is very little difference between the sound received by the left and right ears, and the difference of the received sound at the two ears is what the brain uses to calculate the direction of a sound.

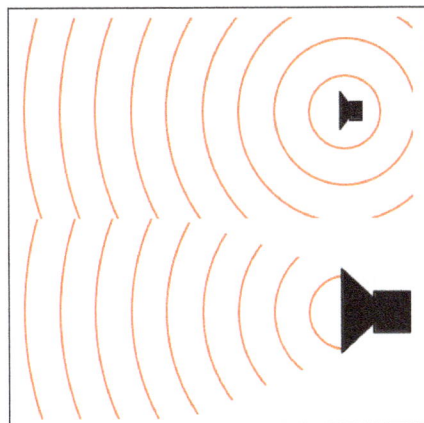

Size matters when designing loudspeakers. Small speakers put out sound in all directions, while large speakers broadcast in a cone in front of them.

Even the way that sound is generated is affected by wavelength. For loudspeakers that are much smaller than a wavelength, sound will tend to spread evenly in all directions. This is called "omni-directionality." When loudspeakers get larger, they become more directional, with speakers that are very large compared to a wavelength projecting in a cone-like shape in front of themselves. This is a large part of why high-end speaker systems are made up of multiple loudspeakers (called drivers). The size of each driver is chosen for the wavelengths of sound it reproduces best, ensuring that the sound system can cover the whole range from 17 meters to 1.7 centimeters.

Of course, while the wavelength for audible sounds is relatively wide, it is still limited. For sounds beyond the audible frequencies, however, wavelengths can be even more extreme. Ultrasound (sounds above the limit of human hearing at ~20 kHz), has wavelengths so small that their reflections can be used to image the tiny structures inside our bodies or used by bats and dolphins to detect and differentiate between prey objects. Infrasound (generally defined as frequencies below the lower limit of human hearing or 20 Hz), on the other hand, can have wavelengths so long that the sounds need to be measured using massive networks of sensors working in unison.

Infrasound wavelengths can be so long that huge setups are needed to measure them.

The range of scales, from the size of the observable universe to the width of a single neutron is about 40 orders of magnitude. While the range of wavelengths for sound is not nearly so large, the range that it covers is important for humans, and is a range that we experience in our daily lives. As a result, by listening carefully, we can hear how both the "very large" and the "very small" alter the ways sounds move through the world around us. We can hear how a thick pillow absorbs all but the deepest of sounds, while a tiny speaker can only reproduce the highest.

Expect to see wavelength come up again and again throughout this series. It is a fundamental concept with far-reaching implications for everything from what allows a medium to support sound, to what makes an acoustic resonator ring. And the next time you see the term, keep an eye out for phrases like "order of magnitude," "much larger," and "much smaller," because when it comes to wavelength, scale is everything.

Timbre

In music, timbre, also known as tone color or tone quality (from psychoacoustics), is the perceived sound quality of a musical note, sound or tone. Timbre distinguishes different types of sound production, such as choir voices and musical instruments, such as string instruments, wind instruments, and percussion instruments. It also enables listeners to distinguish different instruments in the same category (e.g., an oboe and a clarinet, both woodwind instruments).

The physical characteristics of sound that determine the perception of timbre include spectrum and envelope. Singers and instrumental musicians can change the timbre of the music they are singing/playing by using different singing or playing techniques. For example, a violinist can use different bowing styles or play on different parts of the string to obtain different timbres (e.g., playing sul tasto produces a light, airy timbre, whereas playing sul ponticello produces a harsh, even and aggressive tone). On electric guitar and electric piano, performers can change the timbre using effects units and graphic equalizers.

In simple terms, timbre is what makes a particular musical sound have a different sound from another. For instance, it is the difference in sound between a guitar and a piano playing the same note at the same volume. Both instruments can sound equally tuned in relation to each other as they play the same note, and while playing at the same amplitude level each instrument will still sound distinctively with its own unique tone color. Experienced musicians are able to distinguish between different instruments of the same type based on their varied timbres, even if those instruments are playing notes at the same fundamental pitch and loudness.

Tone quality and *tone color* are synonyms for *timbre*, as well as the "*texture* attributed to a single instrument". However, the word texture can also refer to the type of music, such as multiple, interweaving melody lines versus a singable melody accompanied by subordinate chords. Hermann von Helmholtz used the German *Klangfarbe* (*tone color*), and John Tyndall proposed an English translation, *clangtint*, but both terms were disapproved of by Alexander Ellis, who also discredits *register* and *color* for their pre-existing English meanings. The sound of a musical instrument may be described with words such as *bright, dark, warm, harsh*, and other terms. There are also colors of noise, such as pink and white. In visual representations of sound, timbre corresponds to the shape of the image, while loudness corresponds to brightness; pitch corresponds to the y-shift of the spectrogram.

The Acoustical Society of America (ASA) Acoustical Terminology definition 12.09 of timbre describes it as "that attribute of auditory sensation which enables a listener to judge that two non-identical sounds, similarly presented and having the same loudness and pitch, are dissimilar", adding, "Timbre depends primarily upon the frequency spectrum, although it also depends upon the sound pressure and the temporal characteristics of the sound".

Attributes

Timbre has been called "the psychoacoustician's multidimensional waste-basket category for everything that cannot be labeled pitch or loudness."

Many commentators have attempted to decompose timbre into component attributes. For example, J. F. Schouten describes the "elusive attributes of timbre" as "determined by at least five major acoustic parameters", which Robert Erickson finds, "scaled to the concerns of much contemporary music":

1. Range between tonal and noiselike character.
2. Spectral envelope.

3. Time envelope in terms of rise, duration, and decay (ADSR, which stands for "attack, decay, sustain, release").

4. Changes both of spectral envelope (formant-glide) and fundamental frequency (micro-intonation).

5. Prefix, or onset of a sound, quite dissimilar to the ensuing lasting vibration.

An example of a tonal sound is a musical sound that has a definite pitch, such as pressing a key on a piano; a sound with a noiselike character would be white noise, the sound similar to that produced when a radio is not tuned to a station.

Erickson gives a table of subjective experiences and related physical phenomena based on Schouten's five attributes:

Subjective	Objective
Tonal character, usually pitched	Periodic sound
Noisy, with or without some tonal character, including rustle noise	Noise, including random pulses characterized by the rustle time (the mean interval between pulses)
Coloration	Spectral envelope
Beginning/ending	Physical rise and decay time
Coloration glide or formant glide	Change of spectral envelope
Microintonation	Small change (one up and down) in frequency
Vibrato	Frequency modulation
Tremolo	Amplitude modulation
Attack	Prefix
Final sound	Suffix

Harmonics

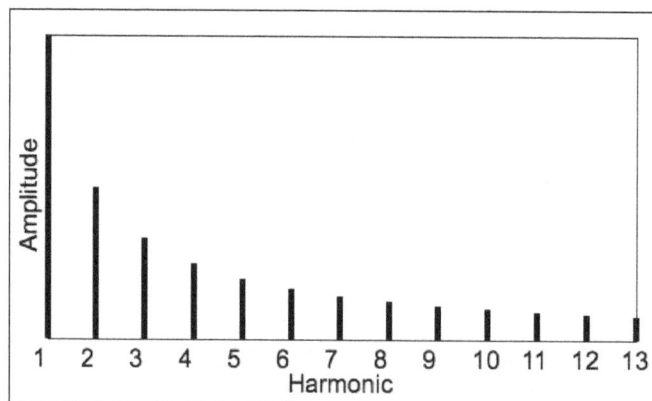

Harmonic spectra.

The richness of a sound or note a musical instrument produces is sometimes described in terms of a sum of a number of distinct frequencies. The lowest frequency is called the *fundamental frequency*, and the pitch it produces is used to name the note, but the fundamental frequency is not always the dominant frequency. The dominant frequency is the frequency that is most heard, and it is always a multiple of the fundamental frequency. For example, the dominant frequency for the transverse flute is double the fundamental frequency. Other significant frequencies are called overtones of the fundamental frequency, which may include harmonics and partials. Harmonics are whole number multiples of the fundamental frequency, such as ×2, ×3, ×4, etc. Partials are other overtones. There are also sometimes subharmonics at whole number *divisions* of the fundamental frequency. Most instruments produce harmonic sounds, but many instruments produce partials and inharmonic tones, such as cymbals and other indefinite-pitched instruments.

When the tuning note in an orchestra or concert band is played, the sound is a combination of 440 Hz, 880 Hz, 1320 Hz, 1760 Hz and so on. Each instrument in the orchestra or concert band produces a different combination of these frequencies, as well as harmonics and overtones. The sound waves of the different frequencies overlap and combine, and the balance of these amplitudes is a major factor in the characteristic sound of each instrument.

William Sethares wrote that just intonation and the western equal tempered scale are related to the harmonic spectra/timbre of many western instruments in an analogous way that the inharmonic timbre of the Thai renat (a xylophone-like instrument) is related to the seven-tone near-equal tempered pelog scale in which they are tuned. Similarly, the inharmonic spectra of Balinese metallophones combined with harmonic instruments such as the stringed rebab or the voice, are related to the five-note near-equal tempered slendro scale commonly found in Indonesian gamelan music.

Envelope

The timbre of a sound is also greatly affected by the following aspects of its *envelope*: attack time and characteristics, decay, sustain, release (ADSR envelope) and transients. Thus these are all common controls on professional synthesizers. For instance, if one takes away the attack from the sound of a piano or trumpet, it becomes more difficult to identify the sound correctly, since the sound of the hammer hitting the strings or the first blast of the player's lips on the trumpet mouthpiece are highly characteristic of those instruments. The envelope is the overall amplitude structure of a sound, so called because the sound just "fits" inside its envelope: what this means should be clear from a time-domain display of almost any interesting sound, zoomed out enough that the entire waveform is visible.

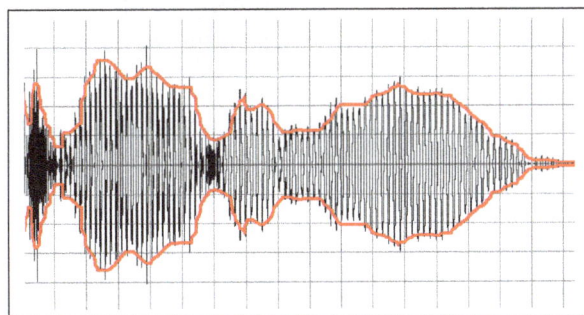

A signal and its envelope marked with red

Psychoacoustic Evidence

Often, listeners can identify an instrument, even at different pitches and loudness, in different environments, and with different players. In the case of the clarinet, acoustic analysis shows waveforms irregular enough to suggest three instruments rather than one. David Luce suggests that this implies that "Certain strong regularities in the acoustic waveform of the above instruments must exist which are invariant with respect to the above variables." However, Robert Erickson argues that there are few regularities and they do not explain our "powers of recognition and identification." He suggests borrowing the concept of subjective constancy from studies of vision and visual perception.

Psychoacoustic experiments from the 1960s onwards tried to elucidate the nature of timbre. One method involves playing pairs of sounds to listeners, then using a multidimensional scaling algorithm to aggregate their dissimilarity judgments into a timbre space. The most consistent outcomes from such experiments are that brightness or spectral energy distribution and the *bite*, or rate and synchronicity and rise time, of the attack are important factors.

Tristimulus Timbre Model

The concept of tristimulus originates in the world of color, describing the way three primary colors can be mixed together to create a given color. By analogy, the musical tristimulus measures the mixture of harmonics in a given sound, grouped into three sections. It is basically a proposal of reducing a huge number of sound partials, that can amount to dozens or hundreds in some cases, down to only three values. The first tristimulus measures the relative weight of the first harmonic; the second tristimulus measures the relative weight of the second, third, and fourth harmonics taken together; and the third tristimulus measures the relative weight of all the remaining harmonics. More evidences, studies and applications would be needed regarding this type of representation, in order to validate it.

$$T_1 = \frac{a_1}{\sum_{h=1}^{H} a_h} \qquad T_2 = \frac{a_2 + a_3 + a_4}{\sum_{h=1}^{H} a_h} \qquad T_3 = \frac{\sum_{h=5}^{H} a_h}{\sum_{h=1}^{H} a_h}$$

Brightness

The term "brightness" is also used in discussions of sound timbres, in a rough analogy with visual brightness. Timbre researchers consider brightness to be one of the perceptually strongest distinctions between sounds, and formalize it acoustically as an indication of the amount of high-frequency content in a sound, using a measure such as the spectral centroid.

Sound Frequency

Sound (or audio) frequency is the speed of the sound's vibration which determines the pitch of the sound. Sound is caused by vibrations which transmit through a medium such as air and reach the ear or some other form of detecting device.

It is measured as the number of wave cycles that occur in one second, with the standard unit of measurement being Hertz (Hz). Sound intensity is measured in Decibels (dB). This is a logarithmic scale in which an increase of 10 dB gives an apparent doubling of loudness.

A frequency of 1 Hz refers to one wave cycle per second, while 20 Hz refers to 20 per second, where the cycles are 20 times shorter and closer together. The audio spectrum is the frequency range which is audible to humans. This generally spans from 20 to 20,000 Hz, although environmental factors influence the precise range for each individual.

Frequencies at the high end of the spectrum are the first to be negatively affected by age and hearing damage as a result of prolonged exposure to loud volumes or noise.As well as intensity and frequency, sound also transmits information. For example, music or speech, transmit information which people may perceive differently from other sounds.

Resistance to the passage of sound defines 'frequency' as:

The number of pressure variations (or cycles) per second that gives a sound its distinctive tone. The unit of frequency is the Hertz (Hz (formerly called 'cycles per second'). A 'frequency band' is a continuous range of frequencies between stated upper and lower limits. An 'octave band' is a frequency band in which the upper limit of the band is twice the frequency of the lower limit. A 'one-third octave band' is a frequency band in which the upper limit of the band is 2 times the frequency of the lower limit.

The sound absorbing characteristics of materials varies significantly with frequency. Low frequency sounds, below 500 Hz, tend to be more difficult to absorb whereas high frequencies sounds, above 500 Hz, are easier to absorb. A material's sound absorbing properties can be expressed by the sound absorption coefficient, alpha, as a function of frequency, where alpha ranges from 0 (total reflection) to 1.00 (total absorption).

Similarly, the sound insulation of materials varies with frequency. Low-frequency sounds tend to be attenuated less by passing through sound insulating materials than high-frequency sounds. As a result, the sound attenuation properties of materials are generally measured at a range of frequencies representative of normal human hearing and this is then compared to a reference frequency profile. Rating of sound insulation in buildings and of building elements. Airborne sound insulation.

Speed of Sound

The speed of sound is the distance travelled per unit time by a sound wave as it propagates through an elastic medium. At 20 °C (68 °F), the speed of sound in air is about 343 metres per second (1,235 km/h; 1,125 ft/s; 767 mph; 667 kn), or a kilometre in 2.9 s or a mile in 4.7 s. It depends strongly on temperature, but also varies by several metres per second, depending on which gases exist in the medium through which a soundwave is propagating.

The speed of sound in an ideal gas depends only on its temperature and composition. The speed has a weak dependence on frequency and pressure in ordinary air, deviating slightly from ideal behavior.

In common everyday speech, *speed of sound* refers to the speed of sound waves in air. However, the speed of sound varies from substance to substance: sound travels most slowly in gases; it travels faster in liquids; and faster still in solids. For example (as noted above), sound travels at 343 m/s in air; it travels at 1,480 m/s in water (4.3 times as fast as in air); and at 5,120 m/s in iron (about 15 times as fast as in air). In an exceptionally stiff material such as diamond, sound travels at 12,000 metres per second (39,000 ft/s)—about 35 times as fast as in air—which is around the maximum speed that sound will travel under normal conditions.

Sound waves in solids are composed of compression waves (just as in gases and liquids), and a different type of sound wave called a shear wave, which occurs only in solids. Shear waves in solids usually travel at different speeds, as exhibited in seismology. The speed of compression waves in solids is determined by the medium's compressibility, shear modulus and density. The speed of shear waves is determined only by the solid material's shear modulus and density.

In fluid dynamics, the speed of sound in a fluid medium (gas or liquid) is used as a relative measure for the speed of an object moving through the medium. The ratio of the speed of an object to the speed of sound in the fluid is called the object's Mach number. Objects moving at speeds greater than *Mach1* are said to be traveling at supersonic speeds.

The transmission of sound can be illustrated by using a model consisting of an array of spherical objects interconnected by springs.

In real material terms, the spheres represent the material's molecules and the springs represent the bonds between them. Sound passes through the system by compressing and expanding the springs, transmitting the acoustic energy to neighboring spheres. This helps transmit the energy in-turn to the neighboring sphere's springs (bonds), and so on.

The speed of sound through the model depends on the stiffness/rigidity of the springs, and the mass of the spheres. As long as the spacing of the spheres remains constant, stiffer springs/bonds transmit energy quicker, while larger spheres transmit the energy slower.

In a real material, the stiffness of the springs is known as the "elastic modulus", and the mass corresponds to the material density. Given that all other things being equal (ceteris paribus), sound will travel slower in spongy materials, and faster in stiffer ones. Effects like dispersion and reflection can also be understood using this model.

For instance, sound will travel 1.59 times faster in nickel than in bronze, due to the greater stiffness of nickel at about the same density. Similarly, sound travels about 1.41 times faster in light hydrogen (protium) gas than in heavy hydrogen (deuterium) gas, since deuterium has similar properties but twice the density. At the same time, "compression-type" sound will travel faster in solids than in liquids, and faster in liquids than in gases, because the solids are more difficult to compress than liquids, while liquids in turn are more difficult to compress than gases.

Some textbooks mistakenly state that the speed of sound increases with density. This notion is illustrated by presenting data for three materials, such as air, water and steel, they each have vastly different compressibility, which more than makes up for the density differences. An illustrative example of the two effects is that sound travels only 4.3 times faster in water than air, despite enormous differences in compressibility of the two media. The reason is that the larger density of water, which works to *slow* sound in water relative to air, nearly makes up for the compressibility differences in the two media.

A practical example can be observed in Edinburgh when the "One o' Clock Gun" is fired at the eastern end of Edinburgh Castle. Standing at the base of the western end of the Castle Rock, the sound of the Gun can be heard through the rock, slightly before it arrives by the air route, partly delayed by the slightly longer route. It is particularly effective if a multi-gun salute such as for "The Queen's Birthday" is being fired.

Compression and Shear Waves

In a gas or liquid, sound consists of compression waves. In solids, waves propagate as two different types. A longitudinal wave is associated with compression and decompression in the direction of travel, and is the same process in gases and liquids, with an analogous compression-type wave in solids. Only compression waves are supported in gases and liquids. An additional type of wave, the transverse wave, also called a shear wave, occurs only in solids because only solids support elastic deformations. It is due to elastic deformation of the medium perpendicular to the direction of wave travel; the direction of shear-deformation is called the "polarization" of this type of wave. In general, transverse waves occur as a pair of orthogonal polarizations.

These different waves (compression waves and the different polarizations of shear waves) may have different speeds at the same frequency. Therefore, they arrive at an observer at different times, an extreme example being an earthquake, where sharp compression waves arrive first and rocking transverse waves seconds later.

The speed of a compression wave in a fluid is determined by the medium's compressibility and density. In solids, the compression waves are analogous to those in fluids, depending on compressibility and density, but with the additional factor of shear modulus which affects compression waves due to off-axis elastic energies which are able to influence effective tension and relaxation in a compression. The speed of shear waves, which can occur only in solids, is determined simply by the solid material's shear modulus and density.

Equations: The speed of sound in mathematical notation is conventionally represented by c, from the Latin *celeritas* meaning "velocity".

For fluids in general, the speed of sound c is given by the Newton–Laplace equation:

$$c = \sqrt{\frac{K_s}{\rho}},$$

where:

- K_s is a coefficient of stiffness, the isentropic bulk modulus (or the modulus of bulk elasticity for gases);

- ρ is the density.

Thus the speed of sound increases with the stiffness (the resistance of an elastic body to deformation by an applied force) of the material and decreases with an increase in density. For ideal gases, the bulk modulus K is simply the gas pressure multiplied by the dimensionless adiabatic index, which is about 1.4 for air under normal conditions of pressure and temperature.

For general equations of state, if classical mechanics is used, the speed of sound c is given by,

$$c = \sqrt{\left(\frac{\partial p}{\partial \rho}\right)_s}$$

where:

- p is the pressure;

- ρ is the density and the derivative is taken isentropically, that is, at constant entropy s.

If relativistic effects are important, the speed of sound is calculated from the relativistic Euler equations.

In a non-dispersive medium, the speed of sound is independent of sound frequency, so the speeds of energy transport and sound propagation are the same for all frequencies. Air, a mixture of oxygen and nitrogen, constitutes a non-dispersive medium. However, air does contain a small amount of CO_2 which is a dispersive medium, and causes dispersion to air at ultrasonic frequencies (> 28 kHz).

In a dispersive medium, the speed of sound is a function of sound frequency, through the dispersion relation. Each frequency component propagates at its own speed, called the phase velocity, while the energy of the disturbance propagates at the group velocity. The same phenomenon occurs with light waves; see optical dispersion for a description.

Dependence on the Properties of the Medium

The speed of sound is variable and depends on the properties of the substance through which the wave is travelling. In solids, the speed of transverse (or shear) waves depends on the shear deformation under shear stress (called the shear modulus), and the density of the medium. Longitudinal (or compression) waves in solids depend on the same two factors with the addition of a dependence on compressibility.

In fluids, only the medium's compressibility and density are the important factors, since fluids do not transmit shear stresses. In heterogeneous fluids, such as a liquid filled with gas bubbles, the density of the liquid and the compressibility of the gas affect the speed of sound in an additive manner, as demonstrated in the hot chocolate effect.

In gases, adiabatic compressibility is directly related to pressure through the heat capacity ratio (adiabatic index), while pressure and density are inversely related to the temperature and molecular weight, thus making only the completely independent properties of *temperature and molecular structure* important (heat capacity ratio may be determined by temperature and molecular structure, but simple molecular weight is not sufficient to determine it).

In low molecular weight gases such as helium, sound propagates faster as compared to heavier gases such as xenon. For monatomic gases, the speed of sound is about 75% of the mean speed that the atoms move in that gas.

For a given ideal gas the molecular composition is fixed, and thus the speed of sound depends only on its temperature. At a constant temperature, the gas pressure has no effect on the speed of sound, since the density will increase, and since pressure and density (also proportional to pressure) have equal but opposite effects on the speed of sound, and the two contributions cancel out exactly. In a similar way, compression waves in solids depend both on compressibility and density—just as in liquids—but in gases the density contributes to the compressibility in such a way that some part of each attribute factors out, leaving only a dependence on temperature, molecular weight, and heat capacity ratio which can be independently derived from temperature and molecular composition. Thus, for a single given gas (assuming the molecular weight does not change) and over a small temperature range (for which the heat capacity is relatively constant), the speed of sound becomes dependent on only the temperature of the gas.

In non-ideal gas behavior regimen, for which the van der Waals gas equation would be used, the proportionality is not exact, and there is a slight dependence of sound velocity on the gas pressure.

Humidity has a small but measurable effect on the speed of sound (causing it to increase by about 0.1%–0.6%), because oxygen and nitrogen molecules of the air are replaced by lighter molecules of water. This is a simple mixing effect.

Altitude Variation and Implications for Atmospheric Acoustics

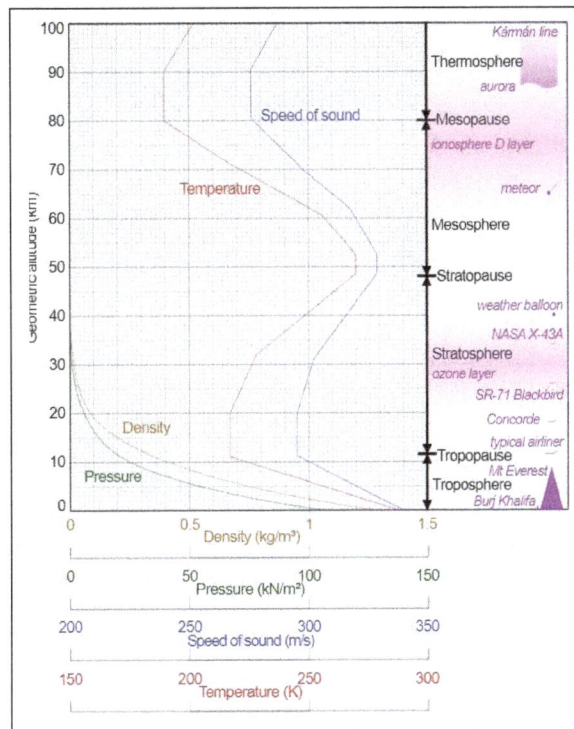

Density and pressure decrease smoothly with altitude, but temperature (red) does not. The speed of sound (blue) depends only on the complicated temperature variation at altitude and can be calculated from it since isolated density and pressure effects on the speed of sound cancel each other. The speed of sound increases with height in two regions of the stratosphere and thermosphere, due to heating effects in these regions.

In the Earth's atmosphere, the chief factor affecting the speed of sound is the temperature. For a given ideal gas with constant heat capacity and composition, the speed of sound is dependent *solely* upon temperature. In such an ideal case, the effects of decreased density and decreased pressure of altitude cancel each other out, save for the residual effect of temperature.

Since temperature (and thus the speed of sound) decreases with increasing altitude up to 11 km, sound is refractedupward, away from listeners on the ground, creating an acoustic shadow at some distance from the source. The decrease of the speed of sound with height is referred to as a negative sound speed gradient.

However, there are variations in this trend above 11 km. In particular, in the stratosphere above about 20 km, the speed of sound increases with height, due to an increase in temperature from heating within the ozone layer. This produces a positive speed of sound gradient in this region. Still another region of positive gradient occurs at very high altitudes, in the aptly-named thermosphere above 90 km.

Practical Formula for Dry Air

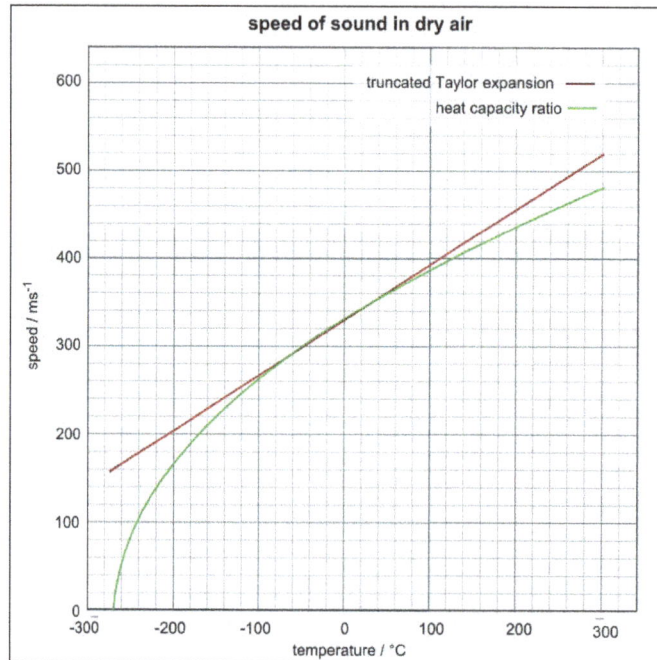

Approximation of the speed of sound in dry air based on the heat capacity ratio (in green) against the truncated Taylor expansion (in red).

The approximate speed of sound in dry (0% humidity) air, in metres per second, at temperatures near 0 °C, can be calculated from:

$$c_{air} = (331.3 + 0.606 \cdot \vartheta) \text{ m/s},$$

where ϑ is the temperature in degrees Celsius (°C).

This equation is derived from the first two terms of the Taylor expansion of the following more accurate equation:

$$c_{air} = 331.3 \sqrt{1 + \frac{\vartheta}{273.15}} \text{ m/s}.$$

Dividing the first part, and multiplying the second part, on the right hand side, by $\sqrt{273.15}$ gives the exactly equivalent form,

$$c_{air} = 20.05 \sqrt{\vartheta + 273.15} \text{ m/s}.$$

which can also be written as,

$$c_{air} = 20.05\sqrt{T/K} \quad \text{m/s}$$

where T denotes the thermodynamic temperature.

The value of 331.3 m/s, which represents the speed at 0 °C (or 273.15 K), is based on theoretical (and some measured) values of the heat capacity ratio, γ, as well as on the fact that at 1 atm real air is very well described by the ideal gas approximation. Commonly found values for the speed of sound at 0 °C may vary from 331.2 to 331.6 due to the assumptions made when it is calculated. If ideal gas γ is assumed to be 7/5 = 1.4 exactly, the 0 °C speed is calculated to be 331.3 m/s, the coefficient used above.

This equation is correct to a much wider temperature range, but still depends on the approximation of heat capacity ratio being independent of temperature, and for this reason will fail, particularly at higher temperatures. It gives good predictions in relatively dry, cold, low-pressure conditions, such as the Earth's stratosphere. The equation fails at extremely low pressures and short wavelengths, due to dependence on the assumption that the wavelength of the sound in the gas is much longer than the average mean free path between gas molecule collisions. A derivation of these equations will be given in the following section.

A graph comparing results of the two equations is at right, using the slightly different value of 331.5 m/s for the speed of sound at 0 °C.

Speed of Sound in Ideal Gases and Air

For an ideal gas, K (the bulk modulus in equations above, equivalent to C, the coefficient of stiffness in solids) is given by,

$$K = \gamma \cdot p,$$

Thus, from the Newton–Laplace equation above, the speed of sound in an ideal gas is given by,

$$c = \sqrt{\gamma \cdot \frac{p}{\rho}},$$

where:

- γ is the adiabatic index also known as the *isentropic expansion factor*. It is the ratio of specific heats of a gas at a constant-pressure to a gas at a constant-volume(C_p / C_v), and arises because a classical sound wave induces an adiabatic compression, in which the heat of the compression does not have enough time to escape the pressure pulse, and thus contributes to the pressure induced by the compression;

- p is the pressure;

- ρ is the density.

Using the ideal gas law to replace p with nRT/V, and replacing ρ with nM/V, the equation for an ideal gas becomes,

$$c_{\text{ideal}} = \sqrt{\gamma \cdot \frac{p}{\rho}} = \sqrt{\frac{\gamma \cdot R \cdot T}{M}} = \sqrt{\frac{\gamma \cdot k \cdot T}{m}},$$

where:

- c_{ideal} is the speed of sound in an ideal gas;

- R (approximately 8.314, 5 J $\cdot$ mol^{-1} $\cdot$ K^{-1}) is the molar gas constant(universal gas constant);

- k is the Boltzmann constant;

- γ (gamma) is the adiabatic index. At room temperature, where thermal energy is fully partitioned into rotation (rotations are fully excited) but quantum effects prevent excitation of vibrational modes, the value is 7/5 = 1.400 for diatomic molecules, according to kinetic theory. Gamma is actually experimentally measured over a range from 1.399,1 to 1.403 at 0 °C, for air. Gamma is exactly 5/3 = 1.6667 for monatomic gases such as noble gases and it is approximately 1.3 for triatomic molecule gases;

- T is the absolute temperature;

- M is the molar mass of the gas. The mean molar mass for dry air is about 0.028,964,5 kg/mol;

- n is the number of moles;

- m is the mass of a single molecule.

This equation applies only when the sound wave is a small perturbation on the ambient condition, and the certain other noted conditions are fulfilled. Calculated values for c_{air} have been found to vary slightly from experimentally determined values.

Newton famously considered the speed of sound before most of the development of thermodynamics and so incorrectly used isothermal calculations instead of adiabatic. His result was missing the factor of γ but was otherwise correct.

Numerical substitution of the above values gives the ideal gas approximation of sound velocity for gases, which is accurate at relatively low gas pressures and densities (for air, this includes standard Earth sea-level conditions). Also, for diatomic gases the use of γ = 1.4000 requires that the gas exists in a temperature range high enough that rotational heat capacity is fully excited (i.e., molecular rotation is fully used as a heat energy "partition" or reservoir); but at the same time the temperature must be low enough that molecular vibrational modes contribute no heat capacity (i.e., insignificant heat goes into vibration, as all vibrational quantum modes above the minimum-energy-mode, have energies too high to be populated by a significant number of molecules at this temperature). For air, these conditions are fulfilled at room temperature, and also temperatures considerably below room temperature.

For air, we introduce the shorthand,

$$R_* = R / M_{\text{air}}.$$

In addition, we switch to the Celsius temperature $\vartheta = T - 273.15$, which is useful to calculate air speed in the region near 0 °C (about 273 kelvin). Then, for dry air,

$$c_{\text{ideal}} = \sqrt{\gamma \cdot R_* \cdot T} = \sqrt{\gamma \cdot R_* \cdot (\vartheta + 273.15)},$$

$$c_{\text{ideal}} = \sqrt{\gamma \cdot R_* \cdot 273.15} \cdot \sqrt{1 + \frac{\vartheta}{273.15}},$$

where ϑ (theta) is the temperature in degrees Celsius(°C).

Substituting numerical values,

$$R = 8.314510 \, \text{J}/(\text{mol} \cdot \text{K})$$

For the molar gas constant in J/mole/Kelvin, and

$$M_{\text{air}} = 0.0289645 \, \text{kg/mol}$$

For the mean molar mass of air, in kg; and using the ideal diatomic gas value of $\gamma = 1.4000$, we have,

$$c_{\text{air}} = 331.3 \sqrt{1 + \frac{\vartheta}{273.15}} \ \text{m/s}.$$

Finally, Taylor expansion of the remaining square root in ϑ yields,

$$c_{\text{air}} = 331.3 \, (1 + \frac{\vartheta}{2 \cdot 273.15}) \ \text{m/s},$$

$$c_{\text{air}} = (331.3 + 0.606 \cdot \vartheta) \ \text{m/s}.$$

The above derivation includes the first two equations given in the "Practical formula for dry air".

Effects due to Wind Shear

The speed of sound varies with temperature. Since temperature and sound velocity normally decrease with increasing altitude, sound is refracted upward, away from listeners on the ground, creating an acoustic shadow at some distance from the source. Wind shear of 4 m/(s · km) can produce refraction equal to a typical temperature lapse rate of 7.5 °C/km. Higher values of wind gradient will refract sound downward toward the surface in the downwind direction, eliminating the acoustic shadow on the downwind side. This will increase the audibility of sounds downwind. This downwind refraction effect occurs because there is a wind gradient; the sound is not being carried along by the wind.

For sound propagation, the exponential variation of wind speed with height can be defined as follows:

$$U(h) = U(0)h^{\zeta},$$

$$\frac{dU}{dH}(h) = \zeta \frac{U(h)}{h},$$

where:

- $U(h)$ is the speed of the wind at height h;

- ζ is the exponential coefficient based on ground surface roughness, typically between 0.08 and 0.52;

- $dU/dH(h)$ is the expected wind gradient at height h.

In the 1862 American Civil War Battle of Iuka, an acoustic shadow, believed to have been enhanced by a northeast wind, kept two divisions of Union soldiers out of the battle, because they could not hear the sounds of battle only 10 km (six miles) downwind.

In the standard atmosphere:

- T_0 is 273.15 K (= 0 °C = 32 °F), giving a theoretical value of 331.3 m/s (= 1086.9 ft/s = 1193 km/h = 741.1 mph = 644.0 kn). Values ranging from 331.3 to 331.6 m/s may be found in reference literature, however;

- T_{20} is 293.15 K (= 20 °C = 68 °F), giving a value of 343.2 m/s (= 1126.0 ft/s = 1236 km/h = 767.8 mph = 667.2 kn);

- T_{25} is 298.15 K (= 25 °C = 77 °F), giving a value of 346.1 m/s (= 1135.6 ft/s = 1246 km/h = 774.3 mph = 672.8 kn).

In fact, assuming an ideal gas, the speed of sound c depends on temperature only, not on the pressure or density (since these change in lockstep for a given temperature and cancel out). Air is almost an ideal gas. The temperature of the air varies with altitude, giving the following variations in the speed of sound using the standard atmosphere—actual conditions may vary.

Effect of temperature on properties of air			
Temperature T (°C)	Speed of sound c (m/s)	Density of air ρ (kg/m³)	Characteristic specific acoustic impedance z_0 (Pa·s/m)
35	351.88	1.1455	403.2
30	349.02	1.1644	406.5
25	346.13	1.1839	409.4
20	343.21	1.2041	413.3
15	340.27	1.2250	416.9
10	337.31	1.2466	420.5
5	334.32	1.2690	424.3
0	331.30	1.2922	428.0
−5	328.25	1.3163	432.1
−10	325.18	1.3413	436.1
−15	322.07	1.3673	440.3
−20	318.94	1.3943	444.6
−25	315.77	1.4224	449.1

Altitude	Temperature	m/s	km/h	mph	kn
Sea level	15 °C (59 °F)	340	1,225	761	661
11,000 m–20,000 m (Cruising altitude of commercial jets, and first supersonic flight)	−57 °C (−70 °F)	295	1,062	660	573
29,000 m (Flight of X-43A)	−48 °C (−53 °F)	301	1,083	673	585

Given normal atmospheric conditions, the temperature, and thus speed of sound, varies with altitude.

Effect of Frequency and Gas Composition

General Physical Considerations

The medium in which a sound wave is travelling does not always respond adiabatically, and as a result, the speed of sound can vary with frequency.

The limitations of the concept of speed of sound due to extreme attenuation are also of concern. The attenuation which exists at sea level for high frequencies applies to successively lower frequencies as atmospheric pressure decreases, or as the mean free path increases. For this reason, the concept of speed of sound (except for frequencies approaching zero) progressively loses its range of applicability at high altitudes. The standard equations for the speed of sound apply with reasonable accuracy only to situations in which the wavelength of the soundwave is considerably longer than the mean free path of molecules in a gas.

The molecular composition of the gas contributes both as the mass (M) of the molecules, and their heat capacities, and so both have an influence on speed of sound. In general, at the same molecular mass, monatomic gases have slightly higher speed of sound (over 9% higher) because they have a higher $\gamma(5/3 = 1.66)$ than diatomics do $(7/5 = 1.4)$. Thus, at the same molecular mass, the speed of sound of a monatomic gas goes up by a factor of:

$$\frac{c_{gas,monatomic}}{c_{gas,diatomic}} = \sqrt{\frac{5/3}{7/5}} = \sqrt{\frac{25}{21}} = 1.091...$$

This gives the 9% difference, and would be a typical ratio for speeds of sound at room temperature in helium vs. deuterium, each with a molecular weight of 4. Sound travels faster in helium than deuterium because adiabatic compression heats helium more since the helium molecules can store heat energy from compression only in translation, but not rotation. Thus helium molecules (monatomic molecules) travel faster in a sound wave and transmit sound faster. (Sound travels at about 70% of the mean molecular speed in gases; the figure is 75% in monatomic gases and 68% in diatomic gases).

Note that in this example we have assumed that temperature is low enough that heat capacities are not influenced by molecular vibration. However, vibrational modes simply cause gammas which decrease toward 1, since vibration modes in a polyatomic gas give the gas additional ways to store heat which do not affect temperature, and thus do not affect molecular velocity and sound velocity. Thus, the effect of higher temperatures and vibrational heat capacity acts to increase the difference between the speed of sound in monatomic vs. polyatomic molecules, with the speed remaining greater in monatomics.

Practical Application to Air

By far the most important factor influencing the speed of sound in air is temperature. The speed is proportional to the square root of the absolute temperature, giving an increase of about 0.6 m/s per degree Celsius. For this reason, the pitch of a musical wind instrument increases as its temperature increases.

The speed of sound is raised by humidity but decreased by carbon dioxide. The difference between 0% and 100% humidity is about 1.5 m/s at standard pressure and temperature, but the size of the humidity effect increases dramatically with temperature. The carbon dioxide content of air is not fixed, due to both carbon pollution and human breath (e.g., in the air blown through wind instruments).

The dependence on frequency and pressure are normally insignificant in practical applications. In dry air, the speed of sound increases by about 0.1 m/s as the frequency rises from 10 Hz to 100 Hz. For audible frequencies above 100 Hz it is relatively constant. Standard values of the speed of sound are quoted in the limit of low frequencies, where the wavelength is large compared to the mean free path.

As shown above, the approximate value 1000/3 = 333.33 m/s is exact a little below 5 °C and is a good approximation for all "usual" outside temperatures (in temperate climates, at least), hence the usual rule of thumb to determine how far lightning has struck: count the seconds from the start of the lightning flash to the start of the corresponding roll of thunder and divide by 3: the result is the distance in kilometers to the nearest point of the lightning bolt.

Mach Number

U.S. Navy F/A-18 traveling near the speed of sound. The white halo consists of condensed water droplets formed by the sudden drop in air pressure behind the shock cone around the aircraft.

Mach number, a useful quantity in aerodynamics, is the ratio of air speed to the local speed of sound. At altitude, for reasons explained, Mach number is a function of temperature. Aircraft flight instruments, however, operate using pressure differential to compute Mach number, not temperature. The assumption is that a particular pressure represents a particular altitude and, therefore, a standard temperature. Aircraft flight instruments need to operate this way because the stagnation pressure sensed by a Pitot tube is dependent on altitude as well as speed.

Experimental Methods

A range of different methods exist for the measurement of sound in air. The earliest reasonably accurate estimate of the speed of sound in air was made by William Derham and acknowledged by Isaac Newton. Derham had a telescope at the top of the tower of the Church of St Laurence in Upminster, England. On a calm day, a synchronized pocket watch would be given to an assistant who would fire a shotgun at a pre-determined time from a conspicuous point some miles away, across the countryside. This could be confirmed by telescope. He then measured the interval between seeing gunsmoke and arrival of the sound using a half-second pendulum. The distance from where the gun was fired was found by triangulation, and simple division (distance/time) provided velocity. Lastly, by making many observations, using a range of different distances, the inaccuracy of the half-second pendulum could be averaged out, giving his final estimate of the speed of sound. Modern stopwatches enable this method to be used today over distances as short as 200–400 metres, and not needing something as loud as a shotgun.

Single-shot Timing Methods

The simplest concept is the measurement made using two microphones and a fast recording device such as a digital storage scope. This method uses the following idea.

If a sound source and two microphones are arranged in a straight line, with the sound source at one end, then the following can be measured:

1. The distance between the microphones (x), called microphone basis.

2. The time of arrival between the signals (delay) reaching the different microphones (t).

Then $v = x/t$.

Other Methods

In these methods, the time measurement has been replaced by a measurement of the inverse of time (frequency).

Kundt's tube is an example of an experiment which can be used to measure the speed of sound in a small volume. It has the advantage of being able to measure the speed of sound in any gas. This method uses a powder to make the nodes and antinodes visible to the human eye. This is an example of a compact experimental setup.

A tuning fork can be held near the mouth of a long pipe which is dipping into a barrel of water. In this system it is the case that the pipe can be brought to resonance if the length of the air column in the pipe is equal to $(1 + 2n)\lambda / 4$ where n is an integer. As the antinodal point for the pipe at the open end is slightly outside the mouth of the pipe it is best to find two or more points of resonance and then measure half a wavelength between these.

Here it is the case that $v = f\lambda$.

High-precision Measurements in Air

The effect of impurities can be significant when making high-precision measurements. Chemical desiccants can be used to dry the air, but will, in turn, contaminate the sample. The air can be dried cryogenically, but this has the effect of removing the carbon dioxide as well; therefore many high-precision measurements are performed with air free of carbon dioxide rather than with natural air.

A 2002 review found that a 1963 measurement by Smith and Harlow using a cylindrical resonator gave "the most probable value of the standard speed of sound to date." The experiment was done with air from which the carbon dioxide had been removed, but the result was then corrected for this effect so as to be applicable to real air. The experiments were done at 30 °C but corrected for temperature in order to report them at 0 °C. The result was 331.45 ± 0.01 m/s for dry air at STP, for frequencies from 93 Hz to 1,500 Hz.

Speed of Sound in Solids

Three-dimensional Solids

In a solid, there is a non-zero stiffness both for volumetric deformations and shear deformations. Hence, it is possible to generate sound waves with different velocities dependent on the deformation mode. Sound waves generating volumetric deformations (compression) and shear deformations (shearing) are called pressure waves (longitudinal waves) and shear waves (transverse waves), respectively. In earthquakes, the corresponding seismic waves are called P-waves(primary waves) and S-waves (secondary waves), respectively. The sound velocities of these two types of waves propagating in a homogeneous 3-dimensional solid are respectively given by,

$$c_{\text{solid,p}} = \sqrt{\frac{K + \frac{4}{3}G}{\rho}} = \sqrt{\frac{E(1-v)}{\rho(1+v)(1-2v)}},$$

$$c_{\text{solid,s}} = \sqrt{\frac{G}{\rho}},$$

where:

- K is the bulk modulus of the elastic materials;
- G is the shear modulus of the elastic materials;
- E is the Young's modulus;
- ρ is the density;
- v is Poisson's ratio.

The last quantity is not an independent one, as $E = 3K(1 - 2v)$. Note that the speed of pressure waves depends both on the pressure and shear resistance properties of the material, while the speed of shear waves depends on the shear properties only.

Typically, pressure waves travel faster in materials than do shear waves, and in earthquakes this is the reason that the onset of an earthquake is often preceded by a quick upward-downward shock, before arrival of waves that produce a side-to-side motion. For example, for a typical steel alloy, $K = 170$ GPa, $G = 80$ GPaand $\rho = 7,700$ kg/m³, yielding a compressional speed $c_{solid,p}$ of 6,000 m/s. This is in reasonable agreement with $c_{solid,p}$ measured experimentally at 5,930 m/sfor a (possibly different) type of steel. The shear speed $c_{solid,s}$ is estimated at 3,200 m/s using the same numbers.

One-dimensional Solids

The speed of sound for pressure waves in stiff materials such as metals is sometimes given for "long rods" of the material in question, in which the speed is easier to measure. In rods where their diameter is shorter than a wavelength, the speed of pure pressure waves may be simplified and is given by:

$$c_{solid} = \sqrt{\frac{E}{\rho}},$$

where E is Young's modulus. This is similar to the expression for shear waves, save that Young's modulus replaces the shear modulus. This speed of sound for pressure waves in long rods will always be slightly less than the same speed in homogeneous 3-dimensional solids, and the ratio of the speeds in the two different types of objects depends on Poisson's ratio for the material.

Speed of Sound in Liquids

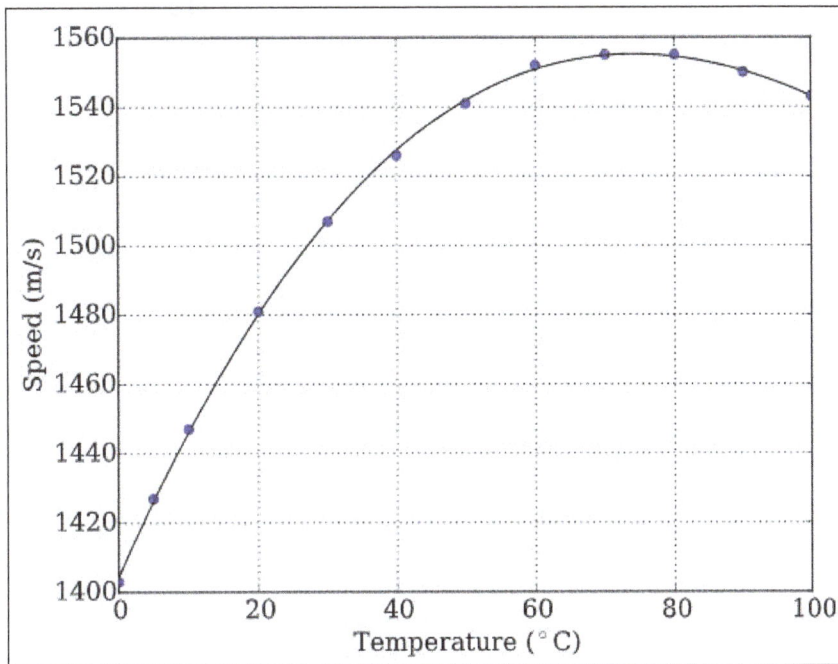

Speed of sound in water vs temperature.

In a fluid, the only non-zero stiffness is to volumetric deformation (a fluid does not sustain shear forces).

Hence the speed of sound in a fluid is given by:

$$c_{fluid} = \sqrt{\frac{K}{\rho}},$$

where K is the bulk modulus of the fluid.

Water

In fresh water, sound travels at about 1481 m/s at 20 °C. Applications of underwater sound can be found in sonar, acoustic communication and acoustical oceanography.

Seawater

In salt water that is free of air bubbles or suspended sediment, sound travels at about 1500 m/s (1500.235 m/s at 1000 kilopascals, 10 °C and 3% salinity by one method). The speed of sound in seawater depends on pressure (hence depth), temperature (a change of 1 °C ~ 4 m/s), and salinity (a change of 1‰ ~ 1 m/s), and empirical equations have been derived to accurately calculate the speed of sound from these variables. Other factors affecting the speed of sound are minor. Since in most ocean regions temperature decreases with depth, the profile of the speed of sound with depth decreases to a minimum at a depth of several hundred metres. Below the minimum, sound speed increases again, as the effect of increasing pressure overcomes the effect of decreasing temperature.

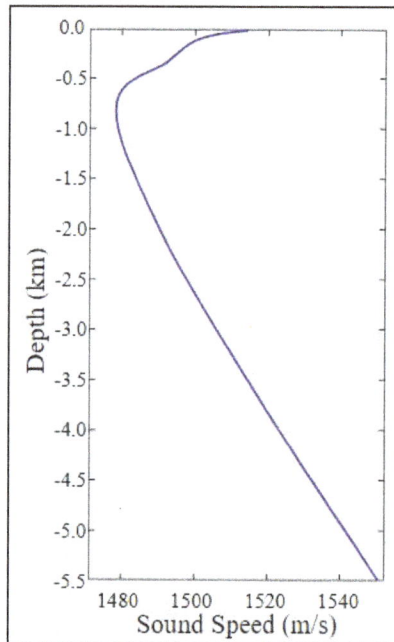

Speed of sound as a function of depth at a position north of Hawaii in the Pacific Ocean derived from the 2005 World Ocean Atlas. The SOFAR channel spans the minimum in the speed of sound at about 750-m depth.

A simple empirical equation for the speed of sound in sea water with reasonable accuracy for the world's oceans is due to Mackenzie:

$$c(T, S, z) = a_1 + a_2 T + a_3 T^2 + a_4 T^3 + a_5(S - 35) + a_6 z + a_7 z^2 + a_8 T(S - 35) + a_9 T z^3,$$

where:

- T is the temperature in degrees Celsius;
- S is the salinity in parts per thousand;
- z is the depth in metres.

The constants $a_1, a_2, ..., a_9$ are:

$$a_1 = 1,448.96, \quad a_2 = 4.591, \quad a_3 = -5.304 \times 10^{-2},$$
$$a_4 = 2.374 \times 10^{-4}, \quad a_5 = 1.340, \quad a_6 = 1.630 \times 10^{-2},$$
$$a_7 = 1.675 \times 10^{-7}, \quad a_8 = -1.025 \times 10^{-2}, \quad a_9 = -7.139 \times 10^{-13},$$

with check value 1550.744 m/s for $T = 25$ °C, $S = 35$ parts per thousand, $z = 1,000$ m. This equation has a standard error of 0.070 m/s for salinity between 25 and 40 ppt. Speed of Sound in Sea-Water for an online calculator.

The Sound Speed vs. Depth graph does *not* correlate directly to the MacKenzie formula. This is due to the fact that the temperature and salinity varies at different depths. When T and S are held constant, the formula itself it always increasing.

Other equations for the speed of sound in sea water are accurate over a wide range of conditions, but are far more complicated, e.g., that by V. A. Del Grossoand the Chen-Millero-Li Equation.

Speed of Sound in Plasma

The speed of sound in a plasma for the common case that the electrons are hotter than the ions (but not too much hotter) is given by the formula:

$$c_s = (\gamma Z k T_e / m_i)^{1/2} = 9.79 \times 10^3 (\gamma Z T_e / \mu)^{1/2} \text{ m/s},$$

where:

- m_i is the ion mass;
- μ is the ratio of ion mass to proton mass $\mu = m_i/m_p$;
- T_e is the electron temperature;
- Z is the charge state;
- k is Boltzmann constant;
- γ is the adiabatic index.

In contrast to a gas, the pressure and the density are provided by separate species, the pressure by the electrons and the density by the ions. The two are coupled through a fluctuating electric field.

Gradients

When sound spreads out evenly in all directions in three dimensions, the intensity drops in proportion to the inverse square of the distance. However, in the ocean, there is a layer called the 'deep sound channel' or SOFAR channel which can confine sound waves at a particular depth.

In the SOFAR channel, the speed of sound is lower than that in the layers above and below. Just as light waves will refract towards a region of higher index, sound waves will refract towards a region where their speed is reduced. The result is that sound gets confined in the layer, much the way light can be confined to a sheet of glass or optical fiber. Thus, the sound is confined in essentially two dimensions. In two dimensions the intensity drops in proportion to only the inverse of the distance. This allows waves to travel much further before being undetectably faint.

A similar effect occurs in the atmosphere. Project Mogul successfully used this effect to detect a nuclear explosion at a considerable distance.

Sound Exposure

Sound exposure is the integral, over time, of squared sound pressure. The SI unit of sound exposure is the pascal squared second (Pa²·s).

Sound exposure, denoted E, is defined by,

$$E = \int_{t_0}^{t_1} p(t)^2 \, dt,$$

where:

- The exposure is being calculated for the time interval between times t_0 and t_1;

- $p(t)$ is the sound pressure at time t, usually A-weighted for sound in air.

Sound exposure level

Sound exposure level (SEL) or acoustic exposure level is a logarithmic measure of the sound exposure of a sound relative to a reference value. Sound exposure level, denoted L_E and measured in dB, is defined by,

$$L_E = \frac{1}{2} \ln\left(\frac{E}{E_0}\right) \mathrm{Np} = \log_{10}\left(\frac{E}{E_0}\right) \mathrm{B} = 10 \log_{10}\left(\frac{E}{E_0}\right) \mathrm{dB},$$

where:

- E is the sound exposure;

- E_0 is the reference sound exposure;

- $1\,\mathrm{Np} = 1$ is the neper;

- $1\,\mathrm{B} = 1/2 \ln 10$ is the bel;

- $1\,\mathrm{dB} = 1/20 \ln 10$ is the decibel.

The commonly used reference sound exposure in air is:

$$E_0 = 400 \, \mu\mathrm{Pa}^2 \cdot \mathrm{s}.$$

The proper notations for sound exposure level using this reference are $L_{W/(400\,\mu\mathrm{Pa}^2 \cdot \mathrm{s})}$ or L_W (re 400 μPa²·s), but the notations dB SEL, dB(SEL), dBSEL, or dB_{SEL} are very common, even if they are not accepted by the SI.

Sound Energy

In physics, sound energy is a form of energy. Sound is a mechanical wave and as such consists physically in oscillatory elastic compression and in oscillatory displacement of a fluid. Therefore, the medium acts as storage for both potential and kinetic energy.

Consequently, the sound energy in a volume of interest is defined as the sum of the potential and kinetic energy densities integrated over that volume:

$$W = W_{potential} + W_{kinetic} = \int_V \frac{p^2}{2\rho_0 c^2} dV + \int_V \frac{\rho v^2}{2} dV,$$

Here:

- V is the volume of interest;
- p is the sound pressure;
- v is the particle velocity;
- ρ_0 is the density of the medium without sound present;
- ρ is the local density of the medium;
- c is the speed of sound.

Sources of Sound Energy

Whenever an object sends out vibrations you can hear, that is, between 20 and 20,000 cycles per second, it produces sound energy. The vibrations can be carried through air, water or solid materials. Mechanical, electrical, or other forms of energy make objects vibrate. When this happens, the energy radiates as sound.

Acoustic Instruments

Pianos, drums, and xylophones are percussive instruments. With these, a hammer strikes an object and makes it vibrate. The piano wire, the drum head and the xylophone bar vibrate in different ways, making waves in the air that we then hear. These instruments also have built-in amplification. The large body of the piano acts as a sounding board, making the vibrating wire louder.

Brass and wind instruments work differently. They set a column of air into resonance, making strong vibrations. The instrument's valves change the resonant frequency, and thus the pitch of the instrument. They usually have a flared opening to achieve natural amplification.

Electronic Instruments

Electrical vibrations are the starting point of sounds from electronic organs and synthesizers. Circuits create a variety of waveshapes that might mimic standard instruments or make totally new sounds. Since wave generation takes place electronically, it's easy to make new sounds with many different effects. It becomes sound, however, only when the electronic signal goes to an amplifier and speakers.

Living Things

Animals and people make sounds with their vocal cords, their mouths and other body parts. Vocal cords vibrate from air pressure, making sound. Insects rapidly rub their legs, wings or other organs to make noise. In the jungle, parrot screeches can carry for miles. Muscles turn chemical energy into mechanical energy. Squeezing and rubbing body parts turn mechanical energy into sound energy.

Machines

In industry, machines make sound in ways similar to musical instruments. However, machines operate at higher speeds and with more power than instruments. They can be designed with sound-absorbing materials to make them quiet, but they're seldom designed to sound pleasant. Loud, rapid impacts of metal on stone make the percussive noise of a jackhammer. Metal parts, rubbing from friction, create the squeal of brakes. Fifty ignitions per second and the clatter of spinning gears make the roar of an engine.

Nature

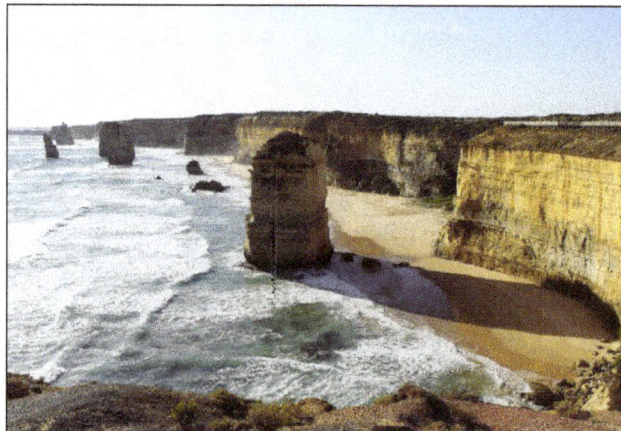

The energy released when falling water hits a beach makes the sound of surf. Lightning explosively heats air, sending sound waves we hear as thunder. Wind, produced by heat from the sun, makes noise through setting objects into vibration. Wind can howl against itself when it gusts.

Sound Energy Density

Sound energy density or sound density is the sound energy per unit volume. The SI unit of sound energy density is the pascal (Pa), that is the joule per cubic metre (J/m³) in SI based units.

Sound energy density, denoted w, is defined by,

$$w = \frac{pv}{c}$$

where:

- p is the sound pressure;
- v is the particle velocity in the direction of propagation;
- c is the speed of sound.

The terms instantaneous energy density, maximum energy density, and peak energy density have meanings analogous to the related terms used for sound pressure. In speaking of average energy density, it is necessary to distinguish between the space average (at a given instant) and the time average (at a given point).

Sound Energy Density Level

The sound energy density level gives the ratio of a sound incidence as a sound energy value in comparison to a reference level of 0 dB (DIN 45630). It is a logarithmic measure of the ratio of two sound energy densities.

The energy produced by vibrations is known as sound,

$$L(E) = 10\log_{10}\left(\frac{E_1}{E_0}\right) \text{dB},$$

where E_1 and E_0 are the energy densities. The unit of the sound energy density level is the decibel (dB).

If E_0 is the standard reference sound energy density of:

$$E_0 = 10^{-12}\,\frac{\text{J}}{\text{m}^3}.$$

Sound Power

Sound power or acoustic power is the rate at which sound energy is emitted, reflected, transmitted or received, per unit time. It is defined as "through a surface, the product of the sound pressure, and the component of the particle velocity, at a point on the surface in the direction normal to the surface, integrated over that surface." The SI unit of sound power is the watt (W). It relates to the power of the sound force on a surface enclosing a sound source, in air. For a sound source, unlike sound pressure, sound power is neither room-dependent nor distance-dependent. Sound pressure is a property of the field at a point in space, while sound power is a property of a sound source, equal to the total power emitted by that source in all directions. Sound power passing through an area is sometimes called sound flux or acoustic flux through that area.

Sound Power Level L_{WA}

Maximum sound power level (L_{WA}) related to a portable air compressor.

Surface enclosing the source. L_{WA} specifies the power delivered to that surface in decibels relative to one picowatt. Devices (e.g., a vacuum cleaner) often have labeling requirements and maximum amounts they are allowed to produce. The A-weighting scale is used in the calculation as the metric is concerned with the loudness as perceived by the human ear. Measurements in accordance with ISO 3744 are taken at 6 to 12 defined points around the device in a hemi-anechoic space. The test environment can be located indoors or outdoors. The required environment is on hard ground in a large open space or hemi-anechoic chamber (free-field over a reflecting plane).

Change of Level	Loudness Perception	Sound Pressure Effect	Sound Intensity Cause
Decibels	Loudness Gain Factor	Voltage Gain Factor	Power Gain Factor
+ 20 dB	4.000	10.000	100.000
+ 10 dB	2.000 •	3.160	10.000
+ 6 dB	1.516	2.000 •	4.000
+ 3 dB	1.232	1.414	2.000 •
± 0 dB	1.000	1.000	1.000
− 3 dB	0.812	0.707	0.500 •
− 6 dB	0.660	0.500 •	0.250
− 10 dB	0.500 •	0.316	0.100
− 20 dB	0.250	0.100	0.010

Sound power, denoted P, is defined by,

$$P = \mathbf{f} \cdot \mathbf{v} = Ap\mathbf{u} \cdot \mathbf{v} = Apv$$

where:

- $\mathbf{f}$ is the sound force of unit vector $\mathbf{u}$;
- $\mathbf{v}$ is the particle velocity of projection v along $\mathbf{u}$;
- A is the area;
- p is the sound pressure.

In a medium, the sound power is given by,

$$P = \frac{Ap^2}{\rho c} \cos \theta,$$

where:

- A is the area of the surface;
- ρ is the mass density;
- c is the sound velocity;
- θ is the angle between the direction of propagation of the sound and the normal to the surface.
- p is the sound pressure.

For example, a sound at SPL = 85 dB or p = 0.356 Pa in air (ρ = 1.2 kg·m^{-3} and c = 343 m·s^{-1}) through a surface of area A = 1 m^2 normal to the direction of propagation (θ = 0°) has a sound energy flux P = 0.3 mW.

This is the parameter one would be interested in when converting noise back into usable energy, along with any losses in the capturing device.

Relationships with other Quantities

Sound power is related to sound intensity:

$$P = AI,$$

where:

- A is the area;
- I is the sound intensity.

Sound power is related sound energy density:

$$P = Acw,$$

where:

- c is the speed of sound;

- w is the sound energy density.

Sound power level (SWL) or acoustic power level is a logarithmic measure of the power of a sound relative to a reference value. Sound power level, denoted L_W and measured in dB, is defined by,

$$L_W = \frac{1}{2}\ln\left(\frac{P}{P_0}\right)\text{Np} = \log_{10}\left(\frac{P}{P_0}\right)\text{B} = 10\log_{10}\left(\frac{P}{P_0}\right)\text{dB},$$

where:

- P is the sound power;

- P_0 is the *reference sound power*;

- $1\,\text{Np} = 1$ is the neper;

- $1\,\text{B} = 1/2 \ln 10$ is the bel;

- $1\,\text{dB} = 1/20 \ln 10$ is the decibel.

The commonly used reference sound power in air is,

$$P_0 = 1\text{pW}.$$

The proper notations for sound power level using this reference are $L_{W/(1\,\text{pW})}$ or L_W (re 1 pW), but the suffix notations dB SWL, dB(SWL), dBSWL, or dB_{SWL} are very common, even if they are not accepted by the SI.

The reference sound power P_0 is defined as the sound power with the reference sound intensity $I_0 = 1$ pW/m² passing through a surface of area $A_0 = 1$ m²:

$$P_0 = A_0 I_0,$$

hence the reference value $P_0 = 1$ pW.

Relationship with Sound Pressure Level

The generic calculation of sound power from sound pressure is as follows:

$$L_W = L_p + 10\log_{10}\left(\frac{A_S}{A_0}\right)\text{dB},$$

where: A_S defines the area of a surface that wholly encompasses the source. This surface may be any shape, but it must fully enclose the source.

In the case of a sound source located in free field positioned over a reflecting plane (i.e. the ground), in air at ambient temperature, the sound power level at distance r from the sound source is approximately related to sound pressure level (SPL) by,

$$L_W = L_p + 10\log_{10}\left(\frac{2\pi r^2}{A_0}\right)\text{dB},$$

where:

- L_p is the sound pressure level;
- $A_0 = 1\text{ m}^2$;
- $2\pi r^2$, defines the surface area of a hemisphere; and
- r must be sufficient that the hemisphere fully encloses the source.

Derivation of this equation:

$$L_W = \frac{1}{2}\ln\left(\frac{P}{P_0}\right)$$

$$= \frac{1}{2}\ln\left(\frac{AI}{A_0 I_0}\right)$$

$$= \frac{1}{2}\ln\left(\frac{I}{I_0}\right) + \frac{1}{2}\ln\left(\frac{A}{A_0}\right).$$

For a *progressive* spherical wave,

$$z_0 = \frac{p}{v},$$

$A = 4\pi r^2$, (the surface area of sphere)

where z_0 is the characteristic specific acoustic impedance.

Consequently,

$$I = pv = \frac{p^2}{z_0},$$

and since by definition $I_0 = p_0^2/z_0$, where $p_0 = 20$ µPa is the reference sound pressure,

$$L_W = \frac{1}{2}\ln\left(\frac{p^2}{p_0^2}\right) + \frac{1}{2}\ln\left(\frac{4\pi r^2}{A_0}\right)$$

$$= \ln\left(\frac{p}{p_0}\right) + \frac{1}{2}\ln\left(\frac{4\pi r^2}{A_0}\right)$$

$$= L_p + 10\log_{10}\left(\frac{4\pi r^2}{A_0}\right)\text{dB}.$$

The sound power estimated practically does not depend on distance. The sound pressure used in the calculation may be affected by distance due to viscous effects in the propagation of sound unless this is accounted for.

Sound Intensity

Intensity is defined to be the power per unit area carried by a wave. Power is the rate at which energy is transferred by the wave. In equation form, *intensity I* is,

$$I = \frac{P}{A},$$

where P is the power through an area A. The SI unit for I is W/m². The intensity of a sound wave is related to its amplitude squared by the following relationship:

$$I = \frac{(\Delta p)^2}{2\rho v_w}$$

Here Δp is the pressure variation or pressure amplitude (half the difference between the maximum and minimum pressure in the sound wave) in units of pascals (Pa) or N/m². The energy (as kinetic energy $\frac{mv^2}{2}$) of an oscillating element of air due to a traveling sound wave is proportional to its amplitude squared. In this equation, ρ is the density of the material in which the sound wave travels, in units of kg/m³, and v_w is the speed of sound in the medium, in units of m/s. The pressure variation is proportional to the amplitude of the oscillation, and so I varies as $(\Delta p)^2$. This relationship is consistent with the fact that the sound wave is produced by some vibration; the greater its pressure amplitude, the more the air is compressed in the sound it creates.

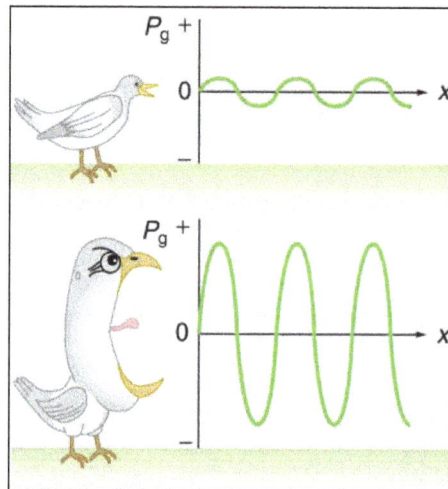

Graphs of the gauge pressures in two sound waves of different intensities. The more intense sound is produced by a source that has larger-amplitude oscillations and has greater pressure maxima and minima. Because pressures are higher in the greater-intensity sound, it can exert larger forces on the objects it encounters.

Sound intensity levels are quoted in decibels (dB) much more often than sound intensities in watts per meter squared. Decibels are the unit of choice in the scientific literature as well as in the popular media. The reasons for this choice of units are related to how we perceive sounds. How our ears perceive sound can be more accurately described by the logarithm of the intensity rather than directly to the intensity. The *sound intensity level β* in decibels of a sound having an intensity *I* in watts per meter squared is defined to be,

$$(\text{dB}) \quad 10\log_{10}\left(\frac{I}{I_o}\right)$$

where $I_0 = 10^{-12}$ W/m² is a reference intensity. In particular, I_0 is the lowest or threshold intensity of sound a person with normal hearing can perceive at a frequency of 1000 Hz. Sound intensity level is not the same as intensity. Because β is defined in terms of a ratio, it is a unitless quantity telling you the *level* of the sound relative to a fixed standard (10^{-12} W/m², in this case). The units of decibels (dB) are used to indicate this ratio is multiplied by 10 in its definition. The bel, upon which the decibel is based, is named for Alexander Graham Bell, the inventor of the telephone.

Sound Intensity Levels and Intensities		
Sound intensity level β (dB)	Intensity I(W/m²)	Example/effect
0	1×10^{-12}	Threshold of hearing at 1000 Hz
10	1×10^{-11}	Rustle of leaves
20	1×10^{-10}	Whisper at 1 m distance
30	1×10^{-9}	Quiet home
40	1×10^{-8}	Average home
50	1×10^{-7}	Average office, soft music
60	1×10^{-6}	Normal conversation
70	1×10^{-5}	Noisy office, busy traffic
80	1×10^{-4}	Loud radio, classroom lecture
90	1×10^{-3}	Inside a heavy truck; damage from prolonged exposure
100	1×10^{-2}	Noisy factory, siren at 30 m; damage from 8 h per day exposure
110	1×10^{-1}	Damage from 30 min per day exposure
120	1	Loud rock concert, pneumatic chipper at 2 m; threshold of pain
140	1×10^{2}	Jet airplane at 30 m; severe pain, damage in seconds
160	1×10^{4}	Bursting of eardrums

The decibel level of a sound having the threshold intensity of 10–12 W/m2 is β = 0 dB, because log101 = 0. That is, the threshold of hearing is 0 decibels. Table gives levels in decibels and intensities in watts per meter squared for some familiar sounds.

One of the more striking things about the intensities in Table is that the intensity in watts per meter squared is quite small for most sounds. The ear is sensitive to as little as a trillionth of a watt per meter squared—even more impressive when you realize that the area of the eardrum is only about 1 cm2, so that only 10–16 W falls on it at the threshold of hearing! Air molecules in a sound wave of this intensity vibrate over a distance of less than one molecular diameter, and the gauge pressures involved are less than 10–9 atm.

Another impressive feature of the sounds in Table is their numerical range. Sound intensity varies by a factor of 1012 from threshold to a sound that causes damage in seconds. You are unaware of this tremendous range in sound intensity because how your ears respond can be described approximately as the logarithm of intensity. Thus, sound intensity levels in decibels fit your experience better than intensities in watts per meter squared. The decibel scale is also easier to relate to because most people are more accustomed to dealing with numbers such as 0, 53, or 120 than numbers such as $1.00 \times 10-11$.

One more observation readily verified by examining Table or using $I=(\Delta p)22\rho vwI=(\Delta p)22\rho vw$ is that each factor of 10 in intensity corresponds to 10 dB. For example, a 90 dB sound compared with a 60 dB sound is 30 dB greater, or three factors of 10 (that is, 103 times) as intense. Another example is that if one sound is 107 as intense as another, it is 70 dB higher.

Table: Ratios of Intensities and Corresponding Differences in Sound Intensity Levels.

$\dfrac{I_2}{I_1}$	$\beta_2 - \beta_1$
2.0	3.0 dB
5.0	7.0 dB
10.0	10.0 dB

Particle Displacement

Particle displacement or displacement amplitude is a measurement of distance of the movement of a sound particle from its equilibrium position in a medium as it transmits a sound wave. The SI unit of particle displacement is the metre (m). In most cases this is a longitudinal wave of pressure (such as sound), but it can also be a transverse wave, such as the vibration of a taut string. In the case of a sound wave travelling through air, the particle displacement is evident in the oscillations of air molecules with, and against, the direction in which the sound wave is travelling.

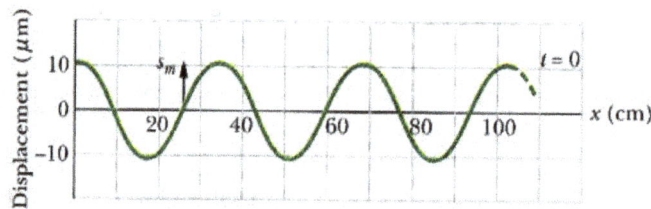

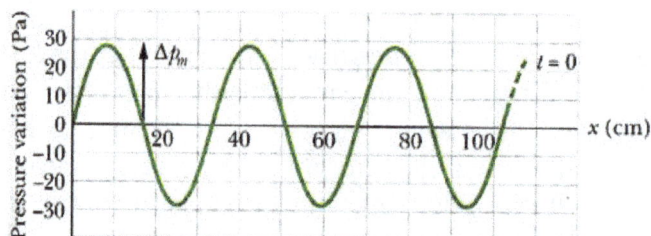

A particle of the medium undergoes displacement according to the particle velocity of the sound wave traveling through the medium, while the sound wave itself moves at the speed of sound, equal to 343 m/s in air at 20 °C.

Particle displacement, denoted δ, is given by,

$$\delta = \int_t v dt \, ,$$

where v is the particle velocity.

Progressive Sine Waves

The particle displacement of a *progressive sine wave* is given by,

$$\delta(\mathbf{r},t) = \delta \cos(\mathbf{k} \cdot \mathbf{r} - \omega t + \varphi_{\delta,0}),$$

where:

- δ is the amplitude of the particle displacement;

- $\varphi_{\delta,0}$ is the phase shift of the particle displacement;

- k is the angular wavevector;

- ω is the angular frequency.

It follows that the particle velocity and the sound pressure along the direction of propagation of the sound wave x are given by,

$$v(\mathbf{r},t) = \frac{\partial \delta(\mathbf{r},t)}{\partial t} = \omega \delta \cos\left(\mathbf{k} \cdot \mathbf{r} - \omega t + \varphi_{\delta,0} + \frac{\pi}{2}\right) = v \cos(\mathbf{k} \cdot \mathbf{r} - \omega t + \varphi_{v,0}),$$

$$p(\mathbf{r},t) = -\rho c^2 \frac{\partial \delta(\mathbf{r},t)}{\partial x} = \rho c^2 k_x \delta \cos\left(\mathbf{k} \cdot \mathbf{r} - \omega t + \varphi_{\delta,0} + \frac{\pi}{2}\right) = p \cos(\mathbf{k} \cdot \mathbf{r} - \omega t + \varphi_{p,0}),$$

where:

- v is the amplitude of the particle velocity;

- $\varphi_{v,0}$ is the phase shift of the particle velocity;

- p is the amplitude of the acoustic pressure;

- $\varphi_{p,0}$ is the phase shift of the acoustic pressure.

Taking the Laplace transforms of v and p with respect to time yields,

$$\hat{v}(\mathbf{r},s) = v \frac{s \cos \varphi_{v,0} - \omega \sin \varphi_{v,0}}{s^2 + \omega^2},$$

$$\hat{p}(\mathbf{r},s) = p\frac{s\cos\varphi_{p,0} - \omega\sin\varphi_{p,0}}{s^2 + \omega^2}.$$

Since $\varphi_{v,0} = \varphi_{p,0}$, the amplitude of the specific acoustic impedance is given by,

$$z(\mathbf{r},s) = |z(\mathbf{r},s)| = \left|\frac{\hat{p}(\mathbf{r},s)}{\hat{v}(\mathbf{r},s)}\right| = \frac{p}{v} = \frac{\rho c^2 k_x}{\omega}.$$

Consequently, the amplitude of the particle displacement is related to those of the particle velocity and the sound pressure by,

$$\delta = \frac{v}{\omega},$$

$$\delta = \frac{p}{\omega z(\mathbf{r},s)}.$$

Particle Velocity

Particle velocity is the velocity of a particle (real or imagined) in a medium as it transmits a wave. The SI unit of particle velocity is the metre per second (m/s). In many cases this is a longitudinal wave of pressure as with sound, but it can also be a transverse wave as with the vibration of a taut string.

When applied to a sound wave through a medium of a fluid like air, particle velocity would be the physical speed of a parcel of fluid as it moves back and forth in the direction the sound wave is travelling as it passes.

Particle velocity should not be confused with the speed of the wave as it passes through the medium, i.e. in the case of a sound wave, particle velocity is not the same as the speed of sound. The wave moves relatively fast, while the particles oscillate around their original position with a relatively small particle velocity. Particle velocity should also not be confused with the velocity of individual molecules.

In applications involving sound, the particle velocity is usually measured using a logarithmic decibel scale called particle velocity level. Mostly pressure sensors (microphones) are used to measure sound pressure which is then propagated to the velocity field using Green's function.

Particle velocity, denoted $\mathbf{v}$, is defined by,

$$\mathbf{v} = \frac{\partial\delta}{\partial t}$$

where δ is the particle displacement.

Progressive Sine Waves

The particle displacement of a *progressive sine wave* is given by,

$$\delta(\mathbf{r},t) = \delta_m \cos(\mathbf{k} \cdot \mathbf{r} - \omega t + \varphi_{\delta,0}),$$

where:

- δ_m is the amplitude of the particle displacement;
- $\varphi_{\delta,0}$ is the phase shift of the particle displacement;
- $\mathbf{k}$ is the angular wavevector;
- ω is the angular frequency.

It follows that the particle velocity and the sound pressure along the direction of propagation of the sound wave x are given by:

$$v(\mathbf{r},t) = \frac{\partial \delta(\mathbf{r},t)}{\partial t} = \omega \delta \cos\left(\mathbf{k} \cdot \mathbf{r} - \omega t + \varphi_{\delta,0} + \frac{\pi}{2}\right) = v_m \cos(\mathbf{k} \cdot \mathbf{r} - \omega t + \varphi_{v,0}),$$

$$p(\mathbf{r},t) = -\rho c^2 \frac{\partial \delta(,)}{\partial} = \rho c^2 k_x \delta \cos\left(\mathbf{k} \cdot \mathbf{r} - \omega t + \varphi_{,0} + \frac{\pi}{}\right) = p_m \cos(\mathbf{k} \cdot \mathbf{r} - \omega t + \varphi_{p,0}),$$

where:

- v_m is the amplitude of the particle velocity;
- $\varphi_{v,0}$ is the phase shift of the particle velocity;
- p_m is the amplitude of the acoustic pressure;
- $\varphi_{p,0}$ is the phase shift of the acoustic pressure.

Taking the Laplace transforms of v and p with respect to time yields,

$$\hat{v}(\mathbf{r},s) = v_m \frac{s \cos \varphi_{v,0} - \omega \sin \varphi_{v,0}}{s^2 + \omega^2},$$

and $$\hat{p}(\mathbf{r},s) = p_m \frac{s \cos \varphi_{p,0} - \omega \sin \varphi_{p,0}}{s^2 + \omega^2}.$$

Since $\varphi_{v,0} = \varphi_{p,0}$, the amplitude of the specific acoustic impedance is given by,

$$z_m(\mathbf{r},s) = |z(\mathbf{r},s)| = \left| \frac{\hat{p}(\mathbf{r},s)}{\hat{v}(\mathbf{r},s)} \right| = \frac{p_m}{v_m} = \frac{\rho c^2 k_x}{\omega}.$$

Consequently, the amplitude of the particle velocity is related to those of the particle displacement and the sound pressure by,

$$v_m = \omega \delta_m,$$

$$v_m = \frac{p_m}{z_m(\mathbf{r},s)}.$$

Particle Velocity Level

Sound velocity level (SVL) or acoustic velocity level or particle velocity level is a logarithmic measure of the effective particle velocity of a sound relative to a reference value. Sound velocity level, denoted L_v and measured in dB, is defined by,

$$L_v = \ln\left(\frac{v}{v_0}\right)\mathrm{Np} = 2\log_{10}\left(\frac{v}{v_0}\right)\mathrm{B} = 20\log_{10}\left(\frac{v}{v_0}\right)\mathrm{dB},$$

where:

- v is the root mean square particle velocity;

- v_0 is the *reference particle velocity*;

- $1\,\mathrm{Np} = 1$ is the neper;

- $1\,\mathrm{B} = 1/2\ln 10$ is the bel;

- $1\,\mathrm{dB} = 1/20\ln 10$ is the decibel.

The commonly used reference particle velocity in air is,

$$v_0 = 5 \times 10^{-8}\ \mathrm{m/s}.$$

The proper notations for sound velocity level using this reference are $L_{v/(5 \times 10^{-8}\ \mathrm{m/s})}$ or L_v (re 5×10^{-8} m/s), but the notations dB SVL, dB(SVL), dBSVL, or $\mathrm{dB_{SVL}}$ are very common, even though they are not accepted by the SI.

Sound Pressure

The Sound Pressure is the force (N) of sound on a surface area (m^2) perpendicular to the direction of the sound.

The SI-units for the sound pressure are N/m^2 or Pa.

Sound is usually measured with microphones responding proportionally to the sound pressure. The power in a sound wave goes as the square of the pressure. (Similarly, electrical power goes as the square of the voltage.) The log of the square of x is just 2 log x, so this introduces a factor of 2 when we convert to decibels for pressures.

The Sound Pressure Level (Decibels)

The lowest sound pressure possible to hear is approximately *2 10⁻⁵ Pa (20 micro Pascal, 0.02 mPa)* or 2 ten billionths of an atmosphere. The minimum audible level occurs between 3000 and 4000 Hz. For a normal human ear pain is experienced at sound pressures of order 60 Pa or 6 10⁻⁴ *atmospheres.*

It is convenient to express sound pressure with the logarithmic decibel scale related to the lowest human hearable sound - *2 10⁻⁵ Pa or 0 dB.*

Sound Pressure Level can be expressed as:

$$L_p = 10 \, log \, (p^2 / p_{ref}^2)$$

$$= 10 \, log \, (p / p_{ref})^2$$

$$= 20 \, log \, (p / p_{ref})$$

where:

L_p = sound pressure level *(dB)*

p = sound pressure (Pa)

p_{ref} = *2 10⁻⁵* - reference sound pressure *(Pa).*

Doubling sound pressure (in Pa) - increases sound pressure level (in dB) with 6 dB (or 20 log (2)).

The chart below shows the sound pressure level decibel scale compared to the sound pressure Pascal scale.

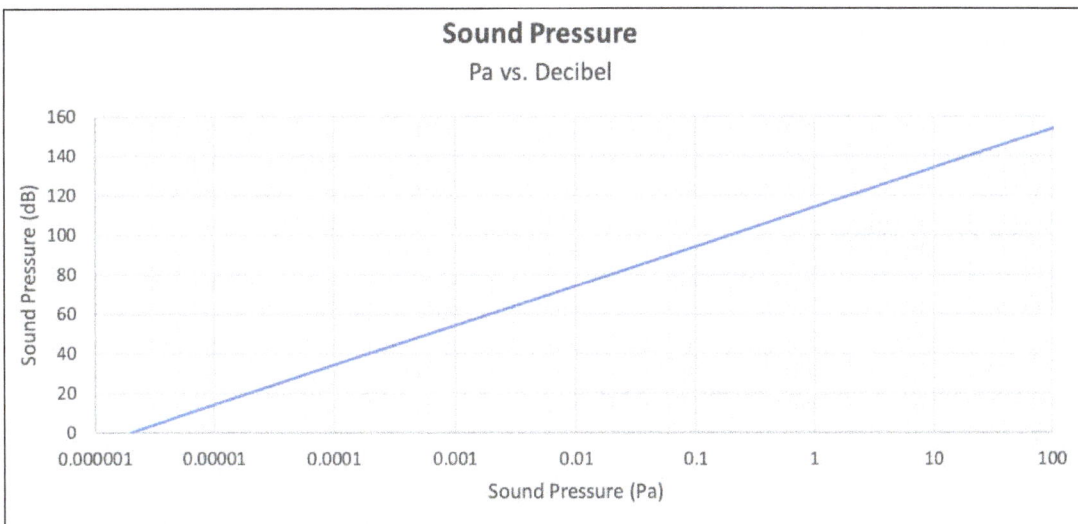

Measuring Sound Pressure

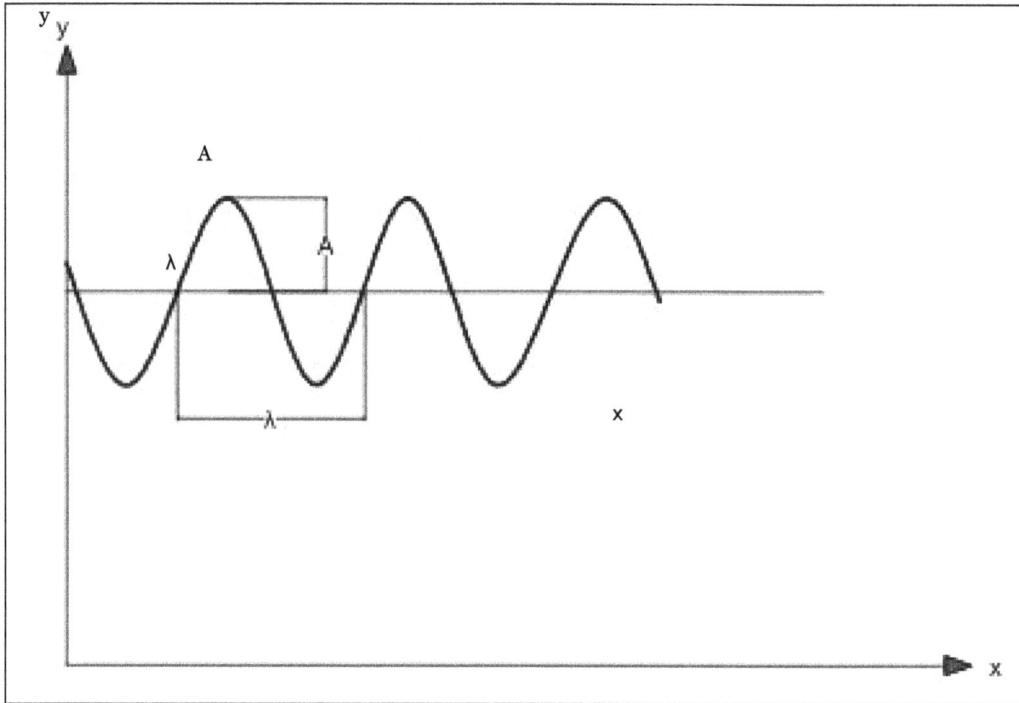

Most Sound Level Meters measures the effective sound pressure which can be expressed as,

$$p_e = p_a / 2^{1/2}$$

where:

p_e = *measured (effective) pressure (Pa).*

p_a = *maximum pressure amplitude in the sound wave (Pa).*

Typical subjective description of sound pressure level:

- *0 - 40 dB* : quiet to very quiet,

- *60 - 80 dB* : noisy,

- *100 dB* : very noisy,

- *> 120 dB* : intolerable.

References

- Amplitude-physics, science: britannica.com, Retrieved 16 may 2019

- Wavelength: acousticstoday.org, Retrieved 16 June 2019

- Cozens, Peter (2006). The Darkest Days of the War: the Battles of Iuka and Corinth. Chapel Hill: The University of North Carolina Press. ISBN 978-0-8078-5783-0

- Wong, George S. K.; Zhu, Shi-ming (1995). "Speed of sound in seawater as a function of salinity, temperature, and pressure". The Journal of the Acoustical Society of America. 97 (3): 1732. Bibcode:1995ASAJ...97.1732W. doi:10.1121/1.413048

- Sound-frequency: designingbuildings.co.uk, Retrieved 24 March 2019

- Robinson, Stephen (22 September 2005). "Technical Guides - Speed of Sound in Sea-Water". National Physical Laboratory. Retrieved 7 December 2016

- Müller, G., Möser, M. (2012). Handbook of Engineering Acoustics. Springer. P. 7. ISBN 9783540694601

- Sound-intensity-and-sound-level, physics: lumenlearning.com, Retrieved 10 July, 2019

- Strong, John Donovan & Hayward, Roger (January 2004). Concepts of Classical Optics. Dover Publications. ISBN 978-0-486-43262-5

5

Categorization of Sound on the basis of Frequency

On the basis of its frequency, sound can be broadly categorized into ultrasound, infrasound and audible sound. Ultrasound has a frequency higher while infrasound has a frequency lower than the limit of human hearing. This chapter closely examines these key categories of sound to provide an extensive understanding of the subject.

Ultrasound

Ultrasound is acoustic (sound) energy in the form of waves having a frequency above the human hearing range. The highest frequency that the human ear can detect is approximately 20 thousand cycles per second (20,000 Hz). This is where the sonic range ends, and where the ultrasonic range begins. Ultrasound is used in electronic, navigational, industrial, and security applications. It is also used in medicine to view internal organs of the body.

Ultrasound can be used to locate objects by means similar to the principle by which radar works. High-frequency acoustic waves reflect from objects, even comparatively small ones, because of the short wavelength. The distance to an object can be determined by measuring the delay between the transmission of an ultrasound pulse and the return of the echo. This is the well-known means by which bats navigate in darkness. It is also believed to be used underwater by cetaceans such as dolphins and whales. Ultrasound can be used in sonar systems to determine the depth of the water in a location, to find schools of fish, to locate submarines, and to detect the presence of SCUBA divers.

In an ultrasonic intrusion detection system, a constant, high-frequency acoustic signal is transmitted by a group of transducers. The ultrasound waves flood the protected area. Receiving transducers monitor the ultrasound reflected by objects in the protected zone. If anything moves, it produces a change in the phase of some of the reflected waves. This phase change is detected by sensitive electronic circuits, which send signals to an alarm or dispatch center. Ultrasonic security systems are also popular among automobile owners. These devices detect motion in the immediate vicinity of a vehicle.

In ultrasonic medical imaging, high-frequency acoustic energy is transmitted into the human body using a set of transducers attached to the skin. The ultrasound waves reflect from boundaries between organs and surrounding fluid, and between regions of differing tissue density. This technique has been used to observe the condition and behavior of fetuses prior to birth. It has also been used to locate tumors, and to observe the condition of the human muscles and bones. Ultrasound is used in industry to analyze the uniformity and purity of liquids and solids. It can also be used for cleaning purposes. Subminiature ultrasonic cleaning instruments are used by some dentists during routine examinations.

Infrasound

Infrasound, sometimes referred to as low-frequency sound, is sound that is lower in frequency than 20 Hz or cycles per second, the "normal" limit of human hearing. Hearing becomes gradually less sensitive as frequency decreases, so for humans to perceive infrasound, the sound pressure must be sufficiently high. The ear is the primary organ for sensing infrasound, but at higher intensities it is possible to feel infrasound vibrations in various parts of the body.

The study of such sound waves is sometimes referred to as infrasonics, covering sounds beneath 20 Hz down to 0.1 Hz and rarely to 0.001 Hz. People use this frequency range for monitoring earthquakes, charting rock and petroleum formations below the earth, and also in ballistocardiography and seismocardiography to study the mechanics of the heart.

Infrasound is characterized by an ability to get around obstacles with little dissipation. In music, acoustic waveguide methods, such as a large pipe organ or, for reproduction, exotic loudspeaker designs such as transmission line, rotary woofer, or traditional subwoofer designs can produce low-frequency sounds, including near-infrasound. Subwoofers designed to produce infrasound are capable of sound reproduction an octave or more below that of most commercially available subwoofers, and are often about 10 times the size.

The Allies of World War I first used infrasound to locate artillery. One of the pioneers in infrasonic research was French scientist Vladimir Gavreau. His interest in infrasonic waves first came about in his laboratory during the 1960s, when he and his laboratory assistants experienced shaking laboratory equipment and pain in the eardrums, but his microphones did not detect audible sound. He concluded it was infrasound caused by a large fan and duct system, and soon got to work preparing tests in the laboratories. One of his experiments was an infrasonic whistle, an oversized organ pipe.

Sources

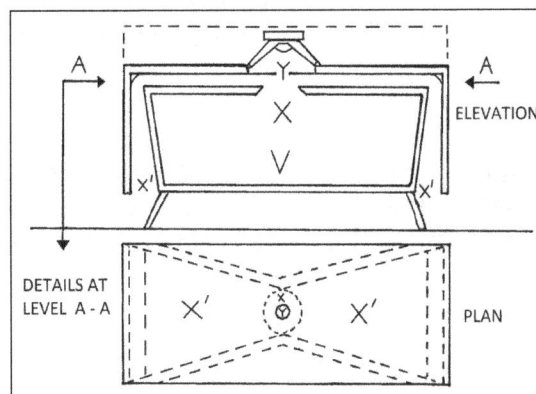

Patent for a double bass reflex loudspeaker enclosure design intended to produce infrasonic frequencies ranging from 5 to 25 hertz, of which traditional subwoofer designs are not readily capable.

Infrasound can result from both natural and man-made sources:

- Natural events: Infrasonic sound sometimes results naturally from severe weather, surf, lee waves, avalanches, earthquakes, volcanoes, bolides, waterfalls, calving of icebergs, aurorae, meteors, lightning and upper-atmospheric lightning. Nonlinear ocean wave interactions in ocean storms produce pervasive infrasound vibrations around 0.2 Hz, known as microbaroms. According to the Infrasonics Program at NOAA, infrasonic arrays can be used to locate avalanches in the Rocky Mountains, and to detect tornadoes on the high plains several minutes before they touch down.

- Animal communication: Whales, elephants, hippopotamuses, rhinoceroses, giraffes, okapis, and alligators are known to use infrasound to communicate over distances—up to hundreds of miles in the case of whales. In particular, the Sumatran rhinoceros has been shown to produce sounds with frequencies as low as 3 Hz which have similarities with the song of the humpback whale. The roar of the tiger contains infrasound of 18 Hz and lower, and the purr of felines is reported to cover a range of 20 to 50 Hz. It has also been suggested that migrating birds use naturally generated infrasound, from sources such as turbulent airflow over mountain ranges, as a navigational aid. Infrasound also may be used for long-distance communication, especially well documented in baleen whales, and African elephants. The frequency of baleen whale sounds can range from 10 Hz to 31 kHz, and that of elephant calls from 15 Hz to 35 Hz. Both can be extremely loud (around 117 dB), allowing communication for many kilometres, with a possible maximum range of around 10 km (6 mi) for elephants, and potentially hundreds or thousands of kilometers for some whales. Elephants also produce infrasound waves that travel through solid ground and are sensed by other herds using their feet, although they may be separated by hundreds of kilometres. These calls may be used to coordinate the movement of herds and allow mating elephants to find each other.

- Human singers: Some vocalists, including Tim Storms, can produce notes in the infrasound range.

- Human created sources: Infrasound can be generated by human processes such as sonic booms and explosions (both chemical and nuclear), or by machinery such as diesel engines, wind turbines and specially designed mechanical transducers (industrial vibration tables). Certain specialized loudspeaker designs are also able to reproduce extremely low frequencies; these include large-scale rotary woofer models of subwoofer loudspeaker, as well as large horn loaded, bass reflex, sealed and transmission line loudspeakers.

Animal Reactions

Some animals have been thought to perceive the infrasonic waves going through the earth, caused by natural disasters, and to use these as an early warning. An example of this is the 2004 Indian Ocean earthquake and tsunami. Animals were reported to have fled the area hours before the actual tsunami hit the shores of Asia. It is not known for sure that this is the cause; some have suggested that it may have been the influence of electromagnetic waves, and not of infrasonic waves, that prompted these animals to flee.

Research in 2013 by Jon Hagstrum of the US Geological Survey suggests that homing pigeons use low-frequency infrasound to navigate.

Human Reactions

20 Hz is considered the normal low-frequency limit of human hearing. When pure sine waves are reproduced under ideal conditions and at very high volume, a human listener will be able to identify tones as low as 12 Hz. Below 10 Hz it is possible to perceive the single cycles of the sound, along with a sensation of pressure at the eardrums.

From about 1000 Hz, the dynamic range of the auditory system decreases with decreasing frequency. This compression is observable in the equal-loudness-level contours, and it implies that even a slight increase in level can change the perceived loudness from barely audible to loud. Combined with the natural spread in thresholds within a population, its effect may be that a very low-frequency sound which is inaudible to some people may be loud to others.

One study has suggested that infrasound may cause feelings of awe or fear in humans. It has also been suggested that since it is not consciously perceived, it may make people feel vaguely that odd or supernatural events are taking place.

A scientist working at Sydney University's Auditory Neuroscience Laboratory reports growing evidence that infrasound may affect some people's nervous system by stimulating the vestibular system, and this has shown in animal models an effect similar to sea sickness.

In research conducted in 2006 focusing on the impact of sound emissions from wind turbines on the nearby population, perceived infrasound has been associated to effects such as annoyance or fatigue, depending on its intensity, with little evidence supporting physiological effects of infrasound below the human perception threshold. Later studies, however, have linked inaudible infrasound to effects such as fullness, pressure or tinnitus, and acknowledged the possibility that it could disturb sleep. Other studies have also suggested associations between noise levels in turbines and self-reported sleep disturbances in the nearby population, while adding that the contribution of infrasound to this effect is still not fully understood.

In a study at Ibaraki University in Japan, researchers said EEG tests showed that the infrasound produced by wind turbines was "considered to be an annoyance to the technicians who work close to a modern large-scale wind turbine."

Jürgen Altmann of the Dortmund University of Technology, an expert on sonic weapons, has said that there is no reliable evidence for nausea and vomiting caused by infrasound.

High volume levels at concerts from subwoofer arrays have been cited as causing lung collapse in individuals who are very close to the subwoofers, especially for smokers who are particularly tall and thin.

In September 2009, London student Tom Reid died of sudden arrhythmic death syndrome (SADS) after complaining that "loud bass notes" were "getting to his heart". The inquest recorded a verdict of natural causes, although some experts commented that the bass could have acted as a trigger.

Air is a very inefficient medium for transferring low frequency vibration from a transducer to the human body. Mechanical connection of the vibration source to the human body, however, provides a potentially dangerous combination. The U.S. space program, worried about the harmful effects of rocket flight on astronauts, ordered vibration tests that used cockpit seats mounted on vibration tables to transfer "brown note" and other frequencies directly to the human subjects. Very high power levels of 160 dB were achieved at frequencies of 2–3 Hz. Test frequencies ranged from 0.5 Hz to 40 Hz. Test subjects suffered motor ataxia, nausea, visual disturbance, degraded task performance and difficulties in communication. These tests are assumed by researchers to be the nucleus of the current urban myth.

The report "A Review of Published Research on Low Frequency Noise and its Effects" contains a long list of research about exposure to high-level infrasound among humans and animals. For instance, in 1972, Borredon exposed 42 young men to tones at 7.5 Hz at 130 dB for 50 minutes. This exposure caused no adverse effects other than reported drowsiness and a slight blood pressure increase. In 1975, Slarve and Johnson exposed four male subjects to infrasound at frequencies from 1 to 20 Hz, for eight minutes at a time, at levels up to 144 dB SPL. There was no evidence of any detrimental effect other than middle ear discomfort. Tests of high-intensity infrasound on animals resulted in measurable changes, such as cell changes and ruptured blood vessel walls.

In February 2005 the television show *MythBusters* used twelve Meyer Sound 700-HP subwoofers—a model and quantity that has been employed for major rock concerts. Normal operating frequency range of the selected subwoofer model was 28 Hz to 150 Hz but the 12 enclosures at *MythBusters* had been specially modified for deeper bass extension. Roger Schwenke and John Meyer directed the Meyer Sound team in devising a special test rig that would produce very high sound levels at infrasonic frequencies. The subwoofers' tuning ports were blocked and their input cards were altered. The modified cabinets were positioned in an open ring configuration: four stacks of three subwoofers each. Test signals were generated by a SIM 3 audio analyzer, with its software modified to produce infrasonic tones. A Brüel & Kjær sound level analyzer, fed with an attenuated signal from a model 4189 measurement microphone, displayed and recorded sound pressure levels. The hosts on the show tried a series of frequencies as low as 5 Hz, attaining a level of 120 decibels of sound pressure at 9 Hz and up to 153 dB at frequencies above 20 Hz, but the rumored physiological effects did not materialize. The test subjects all reported some physical anxiety and shortness of breath, even a small amount of nausea, but this was dismissed by the hosts, noting that sound at that frequency and intensity moves air rapidly in and out of one's lungs. The show declared the brown note myth "busted."

Infrasonic 17 Hz Tone Experiment

On 31 May 2003 a group of UK researchers held a mass experiment, where they exposed some 700 people to music laced with soft 17 Hz sine waves played at a level described as "near the edge of hearing", produced by an extra-long-stroke subwoofer mounted two-thirds of the way from the end of a seven-meter-long plastic sewer pipe. The experimental concert (entitled *Infrasonic*) took place in the Purcell Room over the course of two performances, each consisting of four musical pieces. Two of the pieces in each concert had 17 Hz tones played underneath.

In the second concert, the pieces that were to carry a 17 Hz undertone were swapped so that test results would not focus on any specific musical piece. The participants were not told which pieces included the low-level 17 Hz near-infrasonic tone. The presence of the tone resulted in a significant number (22%) of respondents reporting feeling uneasy or sorrowful, getting chills down the spine or nervous feelings of revulsion or fear.

In presenting the evidence to the British Association for the Advancement of Science, Professor Richard Wiseman said "These results suggest that low frequency sound can cause people to have unusual experiences even though they cannot consciously detect infrasound. Some scientists have suggested that this level of sound may be present at some allegedly haunted sites and so cause people to have odd sensations that they attribute to a ghost—our findings support these ideas."

Suggested Relationship to Ghost Sightings

Psychologist Richard Wiseman of the University of Hertfordshire suggests that the odd sensations that people attribute to ghosts may be caused by infrasonic vibrations. Vic Tandy, experimental officer and part-time lecturer in the school of international studies and law at Coventry University, along with Dr. Tony Lawrence of the University's psychology department, wrote in 1998 a paper called "Ghosts in the Machine" for the Journal of the Society for Psychical Research. Their research suggested that an infrasonic signal of 19 Hz might be responsible for some ghost sightings. Tandy was working late one night alone in a supposedly haunted laboratory at Warwick, when he felt very anxious and could detect a grey blob out of the corner of his eye. When Tandy turned to face the grey blob, there was nothing.

The following day, Tandy was working on his fencing foil, with the handle held in a vice. Although there was nothing touching it, the blade started to vibrate wildly. Further investigation led Tandy to discover that the extractor fan in the lab was emitting a frequency of 18.98 Hz, very close to the resonant frequency of the eye given as 18 Hz by NASA. This, Tandy conjectured, was why he had seen a ghostly figure—it was, he believed, an optical illusion caused by his eyeballs resonating. The room was exactly half a wavelength in length, and the desk was in the centre, thus causing a standing wave which caused the vibration of the foil.

Tandy investigated this phenomenon further and wrote a paper entitled *The Ghost in the Machine*. He carried out a number of investigations at various sites believed to be haunted, including the basement of the Tourist Information Bureau next to Coventry Cathedral and Edinburgh Castle.

Infrasound for Nuclear Detonation Detection

Infrasound is one of several techniques used to identify if a nuclear detonation has occurred. A network of 60 infrasound stations, in addition to seismic and hydroacoustic stations, comprise the International Monitoring System (IMS) that is tasked with monitoring compliance with the Comprehensive Nuclear Test-Ban Treaty (CTBT). IMS Infrasound stations consist of eight microbarometer sensors and space filters arranged in an array covering an area of approximately 1 to 9 km^2. The space filters used are radiating pipes with inlet ports along their length, designed to average out pressure variations like wind turbulence for more precise measurements. The microbarometers used are designed to monitor frequencies below approximately 20 hertz. Sound waves below 20 hertz have longer wavelengths and are not easily absorbed, allowing for detection across large distances.

Infrasound wavelengths can be generated artificially through detonations and other human activity, or naturally from earthquakes, severe weather, lightning, and other sources. Like forensic seismology, algorithms and other filter techniques are required to analyze gathered data and characterize events to determine if a nuclear detonation has actually occurred. Data is transmitted from each station via secure communication links for further analysis. A digital signature is also embedded in the data sent from each station to verify if the data is authentic.

Detection and Measurement

NASA Langley has designed and developed an infrasonic detection system that can be used to make useful infrasound measurements at a location where it was not possible previously. The system comprises an electret condenser microphone PCB Model 377M06, having a 3-inch membrane diameter, and a small, compact windscreen. Electret-based technology offers the lowest possible background noise, because Johnson noise generated in the supporting electronics (preamplifier) is minimized.

The microphone features a high membrane compliance with a large backchamber volume, a prepolarized backplane and a high impedance preamplifier located inside the backchamber. The windscreen, based on the high transmission coefficient of infrasound through matter, is made of a material having a low acoustic impedance and has a sufficiently thick wall to ensure structural stability. Close-cell polyurethane foam has been found to serve the purpose well. In the proposed test, test parameters will be sensitivity, background noise, signal fidelity (harmonic distortion), and temporal stability.

The microphone design differs from that of a conventional audio system in that the peculiar features of infrasound are taken into account. First, infrasound propagates over vast distances through the Earth's atmosphere as a result of very low atmospheric absorption and of refractive ducting that enables propagation by way of multiple bounces between the Earth's surface and the stratosphere. A second property that has received little attention is the great penetration capability of infrasound through solid matter – a property utilized in the design and fabrication of the system windscreens.

Thus the system fulfills several instrumentation requirements advantageous to the application of acoustics: (1) a low-frequency microphone with especially low background noise, which enables detection of low-level signals within a low-frequency passband; (2) a small, compact windscreen that permits (3) rapid deployment of a microphone array in the field. The system also features a data acquisition system that permits real time detection, bearing, and signature of a low-frequency source.

The Comprehensive Nuclear-Test-Ban Treaty Organization Preparatory Commission uses infrasound as one of its monitoring technologies, along with seismic, hydroacoustic, and atmospheric radionuclide monitoring. The loudest infrasound recorded to date by the monitoring system was generated by the 2013 Chelyabinsk meteor.

Audible Sound

The human ear can easily detect frequencies between 20 Hz and 20 KHz. Hence sound waves with frequency ranging from 20 Hz to 20 KHz is known are audible sound. The human ear is sensitive to every minute pressure difference in the air if they are in the audible frequency range. It can detect pressure difference of less than one billionth of atmospheric pressure.

As we grow older and are exposed to sound for longer period of time, our ears get damaged and the upper limit of audible frequencies decreases. For a normal middle-aged adult person, the highest frequency which they can hear clearly is 12-14 kilohertz.

References

- Ultrasound, definition: searchsecurity.techtarget.com, Retrieved 13 July, 2019

- "Gavreau", in Lost Science Archived 19 February 2012 at the Wayback Machine by Gerry Vassilatos. Signals, 1999. ISBN 0-932813-75-5

- Paul Harper (20 February 2013). "Meteor explosion largest infrasound recorded". The New Zealand Herald. APN Holdings NZ. Retrieved 31 March 2013

- Chen, C.H., ed. (2007). Signal and Image Processing for Remote Sensing. Boca Raton: CRC. P. 33. ISBN 978-0-8493-5091-7

- Inaudible-audible-sound, physics: byjus.com, Retrieved 29 July, 2019

- Salt, Alec N.; Kaltenbach, James A. (19 July 2011). "Infrasound From Wind Turbines Could Affect Humans". Bulletin of Science, Technology & Society. 31 (4): 296–302. doi:10.1177/0270467611412555

6
Units of Sound

The acoustic units of sound measurement are known as sound units. The major units of sound include decibel, phon, sone and hertz. The topics elaborated in this chapter will help in gaining a better perspective about these units of sound.

Decibel

The decibel (dB) is a logarithmic unit used to measure sound level. It is also widely used in electronics, signals and communication. The dB is a logarithmic way of describing a ratio. The ratio may be power, sound pressure, voltage or intensity or several other things.

For instance, suppose we have two loudspeakers, the first playing a sound with power P_1, and another playing a louder version of the same sound with power P_2, but everything else (how far away, frequency) kept the same.

Using the decibel unit, the difference in sound level, between the two is defined to be,

$10 \log (P_2/P_1)$ dB   where the log is to base 10.

If the second produces twice as much power than the first, the difference in dB is,

$10 \log (P_2/P_1) = 10 \log 2 = 3$ dB   (to a good approximation).

This is shown on the graph, which plots $10 \log (P_2/P_1)$ against P_2/P_1. To continue the example, if the second had 10 times the power of the first, the difference in dB would be,

$10 \log (P_2/P_1) = 10 \log 10 = 10$ dB.

If the second had a million times the power of the first, the difference in dB would be,

$10 \log (P_2/P_1) = 10 \log 1,000,000 = 60$ dB.

This example shows a feature of decibel scales that is useful in discussing sound: they can describe very big ratios using numbers of modest size. But note that the decibel describes a *ratio*: so far we have not said what power either of the speakers radiates, only the ratio of powers.

Sound pressure, sound level and dB. Sound is usually measured with microphones and they respond proportionally to the sound pressure, p. Now the power in a sound wave, all else equal, goes as the square of the pressure. (Similarly, electrical power in a resistor goes as the square of the voltage.) The log of the x^2 is just 2 log x, so this introduces a factor of 2 when we convert to decibels for pressures. The difference in sound pressure level between two sounds with p_1 and p_2 is therefore:

$$20 \log (p_2/p_1) \text{ dB} = 10 \log (p_2{}^2/p_1{}^2) \text{ dB} = 10 \log (P_2/P_1) \text{ dB} \quad \text{(throughout, the log is to base 10)}.$$

What happens when you have the sound power? The log of 2 is 0.3010, so the log of 1/2 is -0.3, to a good approximation. So, if you halve the power, you reduce the power and the sound level by 3 dB. Halve it again (down to 1/4 of the original power) and you reduce the level by another 3 dB. If you keep on halving the power, you have these ratios.

p,	$\dfrac{p}{\sqrt{2}}$,	$\dfrac{p}{2}$,	$\dfrac{p}{2\sqrt{2}}$,	$\dfrac{p}{4}$,	$\dfrac{p}{4\sqrt{2}}$,	$\dfrac{p}{8}$,	$\dfrac{p}{7\sqrt{2}}$,
I,	$\dfrac{I}{2}$,	$\dfrac{I}{4}$,	$\dfrac{I}{8}$,	$\dfrac{I}{16}$,	$\dfrac{I}{32}$,	$\dfrac{I}{64}$,	$\dfrac{I}{128}$,
L,	L-3dB,	L-6dB,	L-9dB,	L-12dB,	L-15dB,	L-18dB,	L-21dB,

What happens if you add two identical sounds? Do you have double the intensity (increase of 3 dB)? Or do you have double the pressure (increase of 6 dB)? This is a frequently asked question and the answer depends on how you add them.

Sound Files to Show the Size of a Decibel

We saw above that halving the power reduces the sound pressure by $\sqrt{2}$ and the sound level by 3 dB. That is what we have done in the first graphic and sound file below.

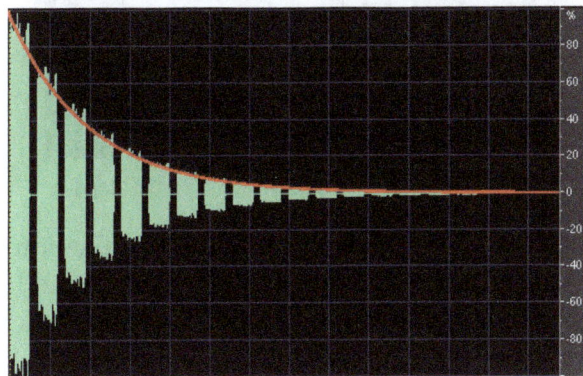

Broadband noise decreasing by 3 dB steps

The first sample of sound is white noise (a mix of all audible frequencies, analogous to white light, which is a mix of all visible frequencies). The second sample is the same noise, with the voltage reduced by a factor of $\sqrt{2}$. Now $1/\sqrt{2}$ is approximately 0.7, so -3 dB corresponds to reducing the voltage or the pressure to 70% of its original value. The green line shows the voltage as a function of time. The red line shows a continuous exponential decay with time. Note that the voltage falls by 50% for every second sample.

A doubling of the power does not make a huge difference to the loudness. We'll discuss this further below, but it's a useful thing to remember when choosing sound reproduction equipment.

How big is a decibel? In the next series, successive samples are reduced by just one decibel.

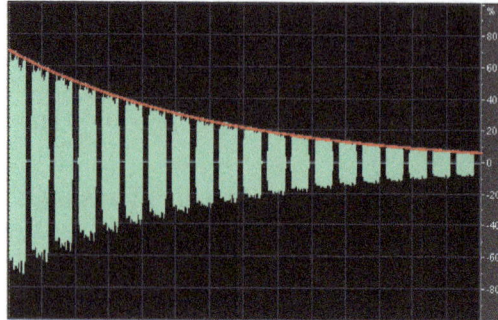
Broadband noise decreasing by 1 dB steps

One decibel is of the same order as the Just Noticeable Difference (JND) for sound level. As you listen to these files, you will notice that the last is quieter than the first, but it is rather less clear to the ear that the second of any pair is quieter than its predecessor. $10*\log_{10}(1.26) = 1$, so to increase the sound level by 1 dB, the power must be increased by 26%, or the voltage by 12%.

What if the difference is less than a decibel? Sound levels are rarely given with decimal places. The reason is that sound levels that differ by less than 1 dB are hard to distinguish, as the next example shows.

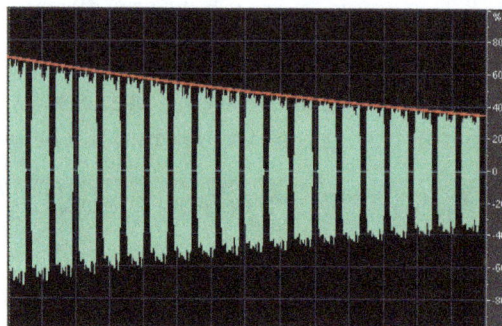
Broadband noise decreasing by 0.3 dB steps

You may notice that the last is quieter than the first, but it is difficult to notice the difference between successive pairs. $10*\log_{10}(1.07) = 0.3$, so to increase the sound level by 0.3 dB, the power must be increased by 7%, or the voltage by 3.5%.

Standard Reference Levels (Absolute Sound Level)

We said above that the decibel is a ratio. So, when it is used to give the sound level for a single sound rather than a ratio, a reference level must be chosen. For sound pressure level, the reference level (for air) is usually chosen as pref = 20 micropascals (20 μPa), or 0.02 mPa. This is very low: it is 2 ten billionths of an atmosphere. Nevertheless, this is about the limit of sensitivity of the human ear, in its sensitive range of frequency. (Usually this sensitivity is only found in rather young people or in people who have not been exposed to loud music or other loud noises. Personal music systems with in-ear speakers are capable of very high sound levels in the ear, and are believed by some to be responsible for much of the hearing loss in young adults in developed countries.)

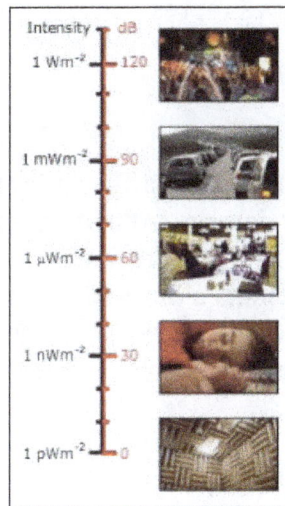

So if you read of a sound pressure level of 86 dB, it means that:

$$20 \log (p_2/p_{ref}) = 86 \text{ dB}$$

where pref is the sound pressure of the reference level, and p2 that of the sound in question. Divide both sides by 20:

$$20 \log (p_2/p_{ref}) = 4.3$$

$$p_2/p_{ref} = 10^{4.3}$$

4 is the log of 10 thousand, 0.3 is the log of 2, so this sound has a sound pressure 20 thousand times greater than that of the reference level ($p_2/p_{ref} = 20{,}000$) or an intensity 400 million times the reference intensity. 86 dB is a loud sound but not dangerous—provided that exposure is brief.

What does 0 dB mean? This level occurs when the measured intensity is equal to the reference level. i.e., it is the sound level corresponding to 0.02 mPa. In this case we have,

$$\text{sound level} = 20 \log (p_{measured}/p_{ref}) = 20 \log 1 = 0 \text{ dB}$$

Remember that decibels measure a ratio. 0 dB occurs when you take the log of a ratio of 1 ($\log 1 = 0$). So 0 dB does not mean no sound, it means a sound level where the sound pressure is equal to that of the reference level. This is a small pressure, but not zero. It is also possible to have negative sound levels: - 20 dB would mean a sound with pressure 10 times smaller than the reference pressure, ie 2 μPa.

Not all sound pressures are equally loud. This is because the human ear does not respond equally to all frequencies: we are much more sensitive to sounds in the frequency range about 1 kHz to 7 kHz (1000 to 7000 vibrations per second) than to very low or high frequency sounds. For this reason, sound meters are usually fitted with a filter whose response to frequency is a bit like that of the human ear. (More about these filters below.) If the "A weighting filter" is used, the sound pressure level is given in units of dB(A) or dBA. Sound pressure level on the dBA scale is easy to measure and is therefore widely used. One reason why it is different from loudness is because the filter does not respond in the same way as the ear. To understand the loudness of a sound, the first thing you need to do consult some curves representing the frequency response of the human ear, given below.

Logarithmic Measures

Why do we use decibels? The ear is capable of hearing a very large range of sounds: the ratio of the sound pressure that causes permanent damage from short exposure to the limit that (undamaged) ears can hear is more than a million. To deal with such a range, logarithmic units are useful: the log of a million is 6, so this ratio represents a difference of 120 dB. Hearing is not inherently logarithmic in response. (Logarithmic measures are also useful when a sound (briefly) increases or decreases exponentially over time. This happens in many applications involving proportional gain or proportional loss).

The Filters used for dBA and dB(C)

The most widely used sound level filter is the A scale, which roughly corresponds to the inverse of the 40 dB (at 1 kHz) equal-loudness curve. Using this filter, the sound level meter is thus less sensitive to very high and very low frequencies. Measurements made on this scale are expressed as dBA. The C scale is practically linear over several octaves and is thus suitable for subjective measurements only for high sound levels. Measurements made on this scale are expressed as dB(C). There is also a (rarely used) B weighting scale, intermediate between A and C.

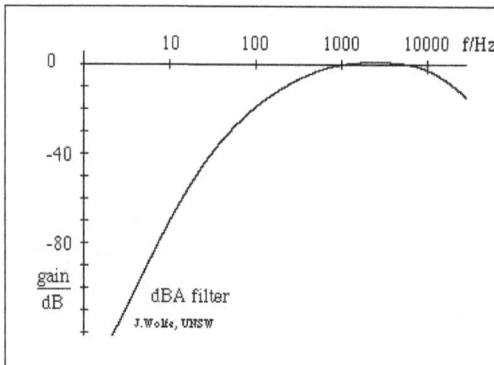

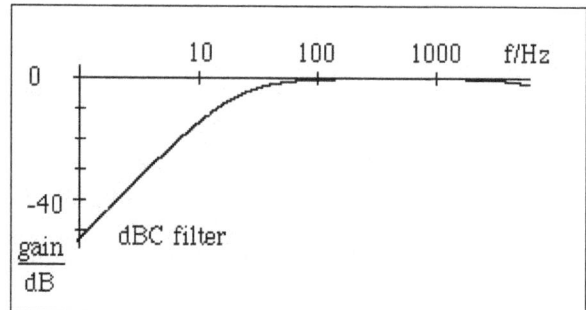

The response of the A filter (left) and C filter, with gains in dB given with respect to 1 kHz.

The equations used to calculate these plots are:

$$A(f) = \frac{12200^2 (f/Hz)^4}{((f/Hz)^2 + 20.6^2)((f/Hz)^2 + 12200^2)\sqrt{(f/Hz)^2 + 107.7^2}\sqrt{(f/Hz)^2 + 737.9^2}}$$

$$C(f) = \frac{12200^2 (f/Hz)^2}{((f/Hz)^2 + 20.6^2)((f/Hz)^2 + 12200^2)}$$

$$\beta_{dBA} = \beta_{dB} + 20\log_{10}\frac{A(f)}{A(1000Hz)} = \beta_{dB} + 20\log_{10}\frac{A(f)}{0.794}$$

$$\beta_{dBA} = \beta_{dB} + 20\log_{10}\frac{C(f)}{C(1000Hz)} = \beta_{dB} + 20\log_{10}\frac{C(f)}{0.993}$$

On the music acoustics and speech acoustics sites, we plot the sound spectra in dB. The reason for this common practice is that the range of measured sound pressures is large.

dB(G) measurements use a narrow band filter that gives high weighting to frequencies between 1 and 20 Hz, and low weighting to others. It thus gives large values for sounds and infrasounds that cannot readily be heard.

Loudness, Phons and Sones and Hearing Response Curves

The phon is a unit that is related to dB by the psychophysically measured frequency response of the ear. At 1 kHz, readings in phons and dB are, by definition, the same. For all other frequencies, the phon scale is determined by the results of experiments in which volunteers were asked to adjust the loudness of a signal at a given frequency until they judged its loudness to equal that of a 1 kHz signal. To convert from dB to phons, you need a graph of such results. Such a graph depends on sound level: it becomes flatter at high sound levels.

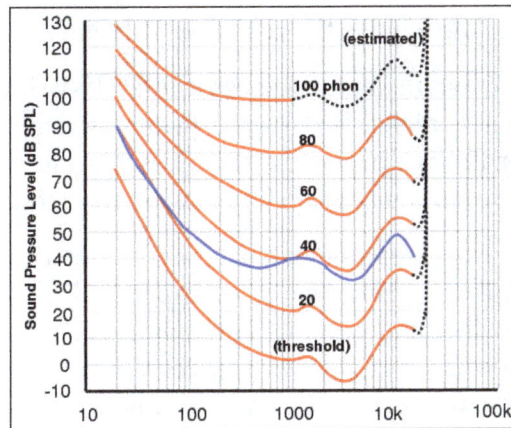

Equal-loudness contours (red) Original ISO standard shown (blue) for 40-phones

This graph, courtesy of Lindosland, shows the 2003 data from the International Standards Organisation for curves of equal loudness determined experimentally. Plots of equal loudness as a function of frequency are often generically called Fletcher-Munson curves.

The sone is derived from psychophysical measurements which involved volunteers adjusting sounds until they judge them to be twice as loud. This allows one to relate perceived loudness to phons. One sone is defined to be equal to 40 phons. Experimentally it was found that, above 40 phons, a 10 dB increase in sound level corresponds approximately to a perceived doubling of loudness. So that approximation is used in the definition of the phon: 1 sone = 40 phon, 2 sone = 50 phon, 4 sone = 60 phon, etc.

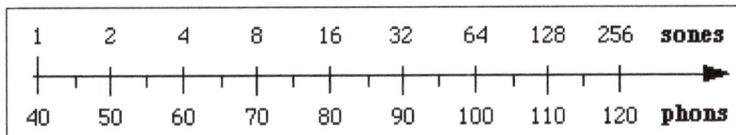

This relation implies that loudness and intensity are related by a power law: loudness in sones is proportional to $(\text{intensity})^{\log 2} = (\text{intensity})^{0.3}$.

Wouldn't it be great to be able to convert from dB (which can be measured by an instrument) to sones (which approximate loudness as perceived by people)? This is usually done using tables that you can find in acoustics handbooks. However, if you don't mind a rather crude approximation, you can say that the A weighting curve approximates the human frequency response at low to moderate sound levels, so dB(A) is very roughly the same as phons, over a limited range of low levels. Then one can use the logarithmic relation between sones and phons described above.

Recording Level and Decibels

Meters measuring recording or output level on audio electronic gear (mixing consoles etc) are almost always recording the AC rms voltage. For a given resistor R, the power P is V^2/R, so,

$$\text{difference in voltage level} = 20 \log\left(V_2/V_1\right) \text{ dB} \;=\; 10 \log\left(V_2^2/V_1^2\right) \text{ dB} \;=\; 10 \log\left(P_2/P_1\right) \text{ dB, or}$$

$$\text{absolute voltage level} = 20 \log\left(V/V_{ref}\right)$$

where V_{ref} is a reference voltage. So what is the reference voltage?

The obvious level to choose is one volt rms, and in this case the level is written as dBV. This is rational, and also convenient with modern analog-digital cards whose maximum range is often about one volt rms. So one has to remember to the keep the level in negative dBV (less than one volt) to avoid clipping the peaks of the signal, but not too negative (so your signal is still much bigger than the background noise).

Sometimes you will see dBm. This used to mean decibels of electrical power, with respect to one milliwatt, and sometimes it still does. However, it's complicated for historical reasons. In the mid twentieth century, many audio lines had a nominal impedance of 600 Ω. If the impedance is purely resisitive, and if you set $V^2/600\ \Omega = 1$ mW, then you get V = 0.775 volts. So, providing you were using a 600 Ω load, 1 mW of power was 0 dBm, which was 0.775 V, so you calibrated your level meters thus. The problem arose because, once a level meter that measures voltage is calibrated like this, it will read 0 dBm at 0.775 V even if it is not connected to 600 Ω So, perhaps illogically, dBm will sometimes mean dB with respect to 0.775 V.

How to convert dBV or dBm into dB of sound level? There is no simple way. It depends on how you convert the electrical power into sound power. Even if your electrical signal is connected directly to a loudspeaker, the conversion will depend on the efficiency and impedance of your loudspeaker. And of course there may be a power amplifier, and various acoustic complications between where you measure the dBV on the mixing desk and where your ears are in the sound field.

Intensity, Radiation and dB

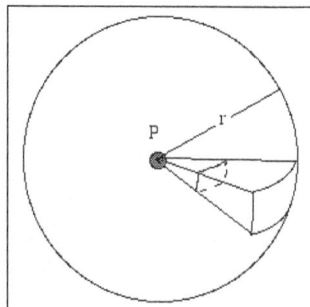

A source that emits radiation equally in all directions is called isotropic. Consider an isolated source of sound, far from any reflecting surfaces – perhaps a bird singing high in the air. Imagine a sphere with radius r, centred on the source. The source outputs a total power P, continuously. This sound power spreads out and is passing through the surface of the sphere. If the source is isotropic, the intensity I is the same everywhere on this surface, by definition. The intensity I is defined as the power per unit area. The surface area of the sphere is 4πr2, so the power (in our example, the sound power) passing through each square metre of surface is, by definition:

$$I = P/4\pi r^2.$$

So we see that, for an isotropic source, intensity is inversely proportional to the square of the distance away from the source:

$$I_2/I_1 = r_1^2/r_2^2.$$

But intensity is proportional to the square of the sound pressure, so we could equally write:

$$I_2/I_1 = r_1^2/r_2^2.$$

So, if we double the distance, we reduce the sound pressure by a factor of 2 and the intensity by a factor of 4: in other words, we reduce the sound level by 6 dB. If we increase r by a factor of 10, we decrease the level by 20 dB, etc.

Be warned, however, that many sources are not isotropic, especially if the wavelength is smaller than, or of a size comparable with the source. Further, reflections are often quite important, especially if the ground is nearby, or if you are indoors.

Pressure, Intensity and Specific Impedance

For acoustic waves, the specific acoustic impedance z is defined as the ratio of the acoustic pressure p to the average particle velocity u, due to the sound ave, z = p/u . In Acoustic impedance, intensity and power, we show how to relate RMS acoustic pressure p and intensity I:

$$I = p^2/z$$

For air, the specific acoustic impedance z is 420 kg.s^{-1}.m^{-2} = 420 Pa.s.m^{-1}. For (fresh) water, the specific acoustic impedance for water is 1.48 MPa.s.m^{-1}. So a sound wave in water with the same pressure has a much lower intensity than one in air.

dBi and Radiation that varies with Direction

Radiation that varies in direction is called anisotropic. For many cases in communication, isotropic radiation is wasteful: why emit a substantial fraction of power upwards if the receiver is, like you, relatively close to ground level. For sound of short wavelength (including most of the important range for speech), a megaphone can help make your voice more anisotropic. For radio, a wide range of designs allows antennae to be highly anisotropic for both transmission and reception.

So, when you interested in emission in (or reception from) a particular direction, you want the ratio of intensity measured in that direction, at a given distance, to be higher than that measured at the same distance from an isotropic radiator (or received by an isotropic receiver). This ratio is called the gain; express the ratio in dB and you have the gain in dBi for that radiator. This unit is mainly used for antennae, either transmitting and receiving, but it is sometimes used for sound sources (and directional microphones).

A few people have written asking for examples in using dB in calculations.

- All else equal, how much louder is loudspeaker driven (in its linear range) by a 100 W amplifier than by a 10 W amplifier?

The powers differ by a factor of ten, which, as we saw above, is 10 dB. All else equal here means that the frequency responses are equal and that the same input signal is used, etc. So the frequency dependence should be the same. 10 dB corresponds to 10 phons. To get a perceived doubling of loudness, you need an increase of 10 phons. So the speaker driven by the 100 W amplifier is twice as loud as when driven by the 10 W, assuming you stay in the linear range and don't distort or destroy the speaker. (The 100 W amplifier produces twice as many sones as does the 10 W.)

- You are standing at a distance R from a small source of sound (size much less than R), at ground level out in the open where reflections may be neglected. The sound level is L. If you now move to a distance nR (n a number, and nR still much greater than the size of the source), what will be the new sound level?

First, note that the neglect of reflections is very important. This calculation will not work inside a room, where reflections from the wall (collectively producing reverberation) make the calculation quite difficult. Out in the open, the sound intensity is proportional to $1/r^2$, where r is the distance from the source. (The constant of proportionality depends on how well the ground reflects, and doesn't concern us here, because it will roughly cancel in the calculation, provided r is reasonably large.) So, if we increase r from R to nR, we decrease the intensity from I to I/n^2.

The difference in decibels between the two signals of intensity I_2 and I_1 is defined above to be,

$$\Delta L \;=\; 10 \,\log\,(I_2/I_1) \;=\; 10 \,\log\,((I/n^2)/I) \;=\; 10 \,\log\,(1/n^2) \;=\; -10 \,\log\,(v) \;=\; -20 \,\log\, n.$$

For example, if n is 2 (ie if we go twice as far away), the intensity is reduced by a factor of four and sound level falls from L to $(L - 6dB)$.

- If, in ideal quiet conditions, a young person can hear a 1 kHz tone at 0 dB emitted by a loudspeaker (perhaps a softspeaker?), by how much must the power of the loudspeaker be increased to raise the sound to 110 dB (a dangerously loud but survivable level)?

The difference in decibels between the two signals of power P2 and P1 is defined above to be,

$$\Delta L \;=\; 10 \,\log\,(P_2/P_1) \; dB \quad \text{so, raising 10 to the power of these two equal quantities:}$$

$$10^{L/10} = P_2/P_1 \quad \text{so:}$$

$$P_2/P_1 = 10^{110/10} = 10^{11} = \text{one hundred thousand million.}$$

which is a demonstration that the human ear has a remarkably large dynamic range, perhaps greater than that of the eye.

- An amplifier has an input of 10 mV and and output of 2 V. What is its voltage gain in dB?

Voltage, like pressure, appears squared in expressions for power or intensity. (The power dissipated in a resistor R is V2/R.) So, by convention, we define:

$$\text{gain} = 20 \log \left(V_{out}/V_{in} \right)$$
$$= 20 \log \left(2V/10mV \right)$$
$$= 46 \text{ dB}$$

We saw that the pressure ratio, expressed in dB, was the same as the power ratio: that was the reason for the factor 20 when defining dB for pressure. It is worth noting that, in the voltage gain example, the power gain of the ampifier is unlikely to equal the voltage gain, which is defined by the convention used here. The power is proportional to the square of the voltage in a given resistor. However, the input and output impedances of amplifiers are often quite different. For instance, a buffer amplifier or emitter follower has a voltage gain of about 1, but a large current gain.

- What is the difference, in dB, between the irradiance (light intensity) on earth (8.3 light minutes from the sun) and on Uranus (160 light minutes)?

Like sound, isotropic light intensity decreases as r^{-2}, so the intensity ratio is $(160/8.3)^2 = 20 \log (160/8.3) = 26$ dB.

Occupational Health and Safety

Different countries and provinces obviously have different laws concerning noise exposure at work, which are enforced with differing enthusiasm. Many such regulations have a limit for exposure to continuous noise of 85 dB(A), for an 8 hour shift. For each 3 dB increase, the allowed exposure is halved. So, if you work in a nightclub where amplified music produces 100 dB(A) near your ears, the allowed exposure is 15 minutes. There is a limit for impulse noise like firearms or tools that use explosive shots. (e.g. 140 dB peak should not be exceeded at any time during the day.) There are many documents providing advice on how to reduce noise exposure at the source (ie turn the music level down), between the source and the ear (i.e. move away from the loudspeakers at a concert) and at the ear (i.e. wear ear plugs or industrial hearing protectors).

Phon

The phon is a unit of loudness level for pure tones. Human sensitivity to sound is variable across different frequencies; therefore, although two different tones may have identical physical intensities, they may be psychoacoustically perceived as differing in loudness. The purpose of the phon is to provide a standard measurement for perceived intensity. The phon is psychophysically matched to a reference frequency of 1 kHz. In other words, the phon matches the sound pressure level (SPL) in decibels of a similarly perceived 1 kHz pure tone . For instance, if a sound is perceived to be equal in intensity to a 1 kHz tone with an SPL of 50 dB, then it has a loudness of 50 phons, regardless of its physical properties. The phon was proposed in DIN 45631 and ISO 532 B by S. S. Stevens.

By definition, the number of phon of a sound is the dB SPL of a sound at a frequency of 1 kHz that sounds just as loud. This implies that 0 phon is the limit of perception, and inaudible sounds have negative phon levels.

The equal-loudness contours are a way of mapping the dB SPL of a pure tone to the perceived loudness level (LN) in phons. These are now defined in the international standard ISO 226:2003, and the research on which this document is based concluded that earlier Fletcher–Munson curves and Robinson–Dadson curves were in error.

The phon unit is not an SI unit in metrology. It is used as a unit of loudness level by the American National Standards Institute (ANSI).

The phon model can be extended with a time-varying transient model which accounts for "turn-on" (initial transient) and long-term listener fatigue effects. This time-varying behavior is the result of psychological and physiological audio processing. The equal-loudness contours on which the phon is based apply only to the perception of pure steady tones; tests using octave or third-octave bands of noise reveal a different set of curves, owing to the way in which the critical bands of our hearing integrate power over varying bandwidths and our brain sums the various critical bands.

Sone

The sone is a unit of loudness, how loud a sound is perceived. Doubling the perceived loudness doubles the sone value. Proposed by Stanley Smith Stevens in 1936, it is not an SI unit.

In acoustics, loudness is the subjective perception of sound pressure. The study of apparent loudness is included in the topic of psychoacoustics and employs methods of psychophysics.

Example Values

Description	Sound pressure	Sound pressure level	Loudness
	Pascal	dB re 20 µPa	Sone
Threshold of pain	~ 100	~ 134	~ 676
Hearing damage during short-term effect	~ 20	~ 120	~ 256
Jet, 100 m away	6 … 200	110 … 140	128 … 1024
Jack hammer, 1 m away / nightclub	~ 2	~ 100	~ 64
Hearing damage during long-term effect	~ $6×10^{-1}$	~ 90	~ 32
Major road, 10 m away	$2×10^{-1}$ … $6×10^{-1}$	80 … 90	16 … 32
Passenger car, 10 m away	$2×10^{-2}$ … $2×10^{-1}$	60 … 80	4 … 16
TV set at home level, 1 m away	~ $2×10^{-2}$	~ 60	~ 4
Normal talking, 1 m away	$2×10^{-3}$ … $2×10^{-2}$	40 … 60	1 … 4

Very calm room	$2\times10^{-4} \dots 6\times10^{-4}$	20 ... 30	0.15 ... 0.4
Rustling leaves, calm breathing	$\sim 6\times10^{-5}$	~ 10	~ 0.02
Auditory threshold at 1 kHz	2×10^{-5}	0	0

Conversion

According to Stevens' definition, a loudness of 1 sone is equivalent to the loudness of a signal at 40 phons, the *loudness level* of a 1 kHz tone at 40 dB SPL. But phons scale with level in dB, not with loudness, so the sone and phon scales are not proportional. Rather, the loudness in sones is, at least very nearly, a power law function of the signal intensity, with an exponent of 0.3. With this exponent, each 10 phon increase (or 10 dB at 1 kHz) produces almost exactly a doubling of the loudness in sones.

Sone	1	2	4	8	16	32	64	128	256	512	1024
Phon	40	50	60	70	80	90	100	110	120	130	140

At frequencies other than 1 kHz, the loudness level in phons is calibrated according to the frequency response of human hearing, via a set of equal-loudness contours, and then the loudness level in phons is mapped to loudness in sones via the same power law.

Loudness N in sones (for $L_N > 40$ phon):

$$N = \left(10^{\frac{L_N-40}{10}}\right)^{0.30103} \approx 2^{\frac{L_N-40}{10}}$$

or loudness level L_N in phons (for $N > 1$ sone):

$$L_N = 40 + 10\log_2(N)$$

Corrections are needed at lower levels, near the threshold of hearing.

These formulas are for single-frequency sine waves or narrowband signals. For multi-component or broadband signals, a more elaborate loudness model is required, accounting for critical bands.

To be fully precise, a measurement in sones must be specified in terms of the optional suffix G, which means that the loudness value is calculated from frequency groups, and by one of the two suffixes D (for direct field or free field) or R (for room field or diffuse field).

Hertz

Hertz, in short Hz, is the basic unit of frequency, to commemorate the discovery of electromagnetic waves by the German physicist Heinrich Rudolf Hertz. In 1888, German physicist Heinrich Rudolf Hertz, the first person confirmed the existence of radio waves, and had a great contribution in Electromagnetism, so the SI unit of frequency Hertz is named as his name.

Uses of Hz

Hz (Hertz) is the frequency unit of the vibration cycle time of electric, magnetic, acoustic and mechanical vibration, i.e. the number of times per second (cycle/sec).

1 Hertz means one vibration cycle per second, 50 Hertz means 50 vibration cycles per second while 60 Hertz means 60 vibration cycles per second. Hz is a very small unit, usually coupled with kHz (kilohertz), MHz (Megahertz), GHz (Gigahertz) etc.

kHz is frequency unit of alternating current (AC) or electromagnetic wave (EM), equal to 1000 hertz (1000 Hz). This unit is also used for measuring and describing the signal bandwidth.

1 kHz frequency AC signal is within the human auditory sensation area. The EM wavelength of 1 kHz signal is 300 km, which is about 190 miles. Standard amplitude modulation (AM) broadcast bandwidth is in the range of 535 kHz to 1605 kHz. Some EM transmissions are in millions of kHz.

kHz is a relative small unit of frequency, more common units are MHz, equal to 1,000,000 Hz or 1,000 kHz, and GHz, which equal to 1,000,000,000 Hz or 1,000,000 kHz.

Hz Common Values

For sounds, the human hearing range is 20 Hz ~ 20000 Hz, lower than this range is called infrasound, higher than this range is called ultrasound.

ITU defines radio frequency range:

1. Ultra Low Frequency (ULF): 3 ~ 30 kilohertz (kHz).
2. Low Frequency (LF): 30 ~ 300 kilohertz (kHz).
3. Intermediate Frequency (MF): 300 ~ 3000 kilohertz (kHz).
4. High Frequency (HF): 3 ~ 30 megahertz (MHz).
5. Very High Frequency (VHF): 30 ~ 300 megahertz (MHz).
6. Ultra High Frequency (UHF): 300 ~ 3000 megahertz (MHz).
7. Super High Frequency (SHF): 3 ~ 30 GHz (GHz).
8. Extremely High Frequency (EHF): 30 ~ 300 GHz (GHz).

What is Hz Converter?

A Hz Converter is an electronic device to convert mains power (50 Hz, 60 Hz, etc.) to variable Hertz, variable Volts for home/industry appliances compatibility. It's different with a variable frequency drive which is only for AC motors due the output waveform is square wave, and output Hertz and Volts can not be changed in separate. A Hz converter outputs pure sine wave, Hz and Volts can be adjusted in separate, e.g. 50 Hz 220V, 50 Hz 400V, 60 Hz 110V, 60 Hz 480V, 400 Hz 115V, 230V, 240V etc. with random combination for different equipment running on its perfect condition. By using a Hz converter, you can even go for much higher frequencies, like 120 Hz, 400 Hz for aircrafts, ships, military utilities etc.

Can a 50 Hertz Motor Run on 60 Hertz Power System

Since the formula for controlling the synchronous speed of a three phase motor is = [(120* Hz) / Motor Poles] if this is a 4-pole motor then at 50 Hz the speed would be 1500 RPM whereas at 60 Hz the speed would be 1800 RPM. Since motors are constant torque machines then by applying the formula that HP = (torque*RPM)/5252 then you can see that with a 20% increase in speed the motor would also then be able to produce 20% more horsepower. The motor would be able to produce rated torque at both frequencies only apply if the V/Hz ratio is constant, meaning that at 50 Hz the supply voltage would need to be 380V and at 60 Hz the supply voltage would need to be 460V. In both cases the V/Hz ratio is 7.6V/Hz.

References

- Brian C. J. Moore (2007). Cochlear hearing loss: physiological, psychological and technical issues (2nd ed.). Wiley-Interscience. pp. 94–95. ISBN 978-0-470-51633-1

- Loudness Units: Phons and Sones". Hyperphysics.phy-astr.gsu.edu. Retrieved 2019-01-12

- Pease, C.B. (1974-07-01). "Combining the sone and phon scales". Applied Acoustics. 7 (3): 167–181. doi:10.1016/0003-682X(74)90011-5. ISSN 0003-682X

- Hugo Fastl and Eberhard Zwicker (2007). Psychoacoustics: facts and models (3rd ed.). Springer. P. 207. ISBN 978-3-540-23159-2

- Stanley Smith Stevens: A scale for the measurement of the psychological magnitude: loudness. See: Psychological Review. 43, Nr. 5,APA Journals, 1936, pp. 405-416

- Hertz: gohz.com, Retrieved 15 July, 2019

7

Sound Propagation: Reflection, Refraction and Diffraction

Sound is a vibration which needs a transmission medium for its propagation. During propagation, sound waves can undergo refraction, reflection or diffraction. This chapter closely examines these key concepts of sound propagation to provide an extensive understanding of the subject.

Propagation of Sound

Sound is a sequence of waves of pressure which propagates through compressible media such as air or water. (Sound can propagate through solids as well, but there are additional modes of propagation). During their propagation, waves can be reflected, refracted, or attentuated by the medium. The purpose of this experiment is to examine what effect the characteristics of the medium have on sound.

All media have three properties which affect the behavior of sound propagation:

1. A relationship between density and pressure. This relationship, affected by temperature, determines the speed of sound within the medium.

2. The motion of the medium itself, e.g., wind. Independent of the motion of sound through the medium, if the medium is moving, the sound is further transported.

3. The viscosity of the medium. This determines the rate at which sound is attenuated. For many media, such as air or water, attenuation due to viscosity is negligible.

What happens when sound is propagating through a medium which does not have constant properties? For example, when sound speed increases with height? Sound waves are refracted. They can be focused or dispersed, thus increasing or decreasing sound levels, precisely as an optical lens increases or decreases light intensity.

One way that the propagation of sound can be represented is by the motion of wavefronts- lines of constant pressure that move with time. Another way is to hypothetically mark a point on a wavefront and follow the trajectory of that point over time. This latter approach is called ray-tracing and shows most clearly how sound is refracted.

In the simulation which follows, the effects of the medium on sound propagation can be visualized. The user can generate a variety of sound-speed profiles and wind-speed profiles by clicking on the profile choices and dragging the red dots to establish amplitudes. Two sound sources are available: a spherical source, in which initial sound waves emanate uniformly in all directions; and a planar source , in which initial sound waves emanate in a single direction. The location of the source and it orientation can be changed by dragging the red dots. Sound propagation in this simulation is in two dimensions; and media profiles depend on height only. Pressing 'Start' will begin the simulation. Propagation is represented both by rays (black) and wavefronts (red). Note that the sound speed Co is artificially low to accentuate the effects of the medium. (Sound speed in air is nominally 340m/s; in water, 1500m/s). Data, including sound speed, wind speed, and derivatives, may be obtained by clicking anywhere within the orange propagation field.

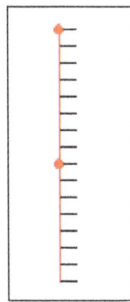

Reflection of Sound

Sound wave also gets reflected as light waves do. Bouncing back of sound wave from the surface of solid or liquid is called reflection of sound. Reflection of sound follows the Laws of Reflection as light wave does. This means the angle of incident wave and reflected wave to the normal are equal. For reflection of sound a polished or rough and big obstacle is necessary.

Use of Reflection of Sound

Reflection of sound is used in many devices. For example; megaphone, loudspeaker, bulb horn, stethoscope, hearing aid, sound board etc.

Loudspeaker, Megaphone and Bulb Horn

Loudspeaker, Megaphone and bulb horn are devices used to send the sound in desired direction without spreading the sound all around. These devices act on the laws of reflection of sound wave.

In such devices, a funnel like cone shaped tube is used. Sound is introduced at the narrower end of tube and let to come out from wider end. Because of successive reflections, the amplitude of sound is added up which makes the sound louder. The name 'Loudspeaker' is given as it is used to make the sound louder.

Stethoscope

Stethoscope is used to hear the sounds of internal organs of a patient; for diagnostic purposes. It works on the laws of reflection of sound.

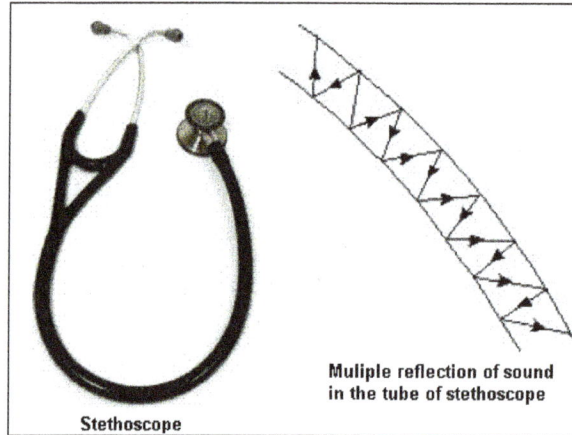

Mulitple reflection of sound in the tube of stethoscope

Stethoscope

In stethoscope, sound is received by chest piece and sent to the earpieces by multiple reflecting through a long tube. Doctors diagnose the condition of an organ of the human body by hearing the sound using the stethoscope. Stethoscope has become the symbol of the medical profession since its invention.

Soundboard

Sound board is used to send the sound towards audience in big hall or auditorium. This works on the basis of laws of reflection of sound waves. Sound board is a big concave board and is set in such a fashion behind the stage that speaker is at the focus. Sound coming from speaker falls over sound board and gets reflected towards the audience. As a result, the audience sitting in the hall even at far distance from the speaker can clearly hear what the speaker is saying. Additionally, the ceiling of the auditorium is also made curved so that it also acts like sound board. The curved surface of the ceiling reflects the sound waves and facilitates better hearing.

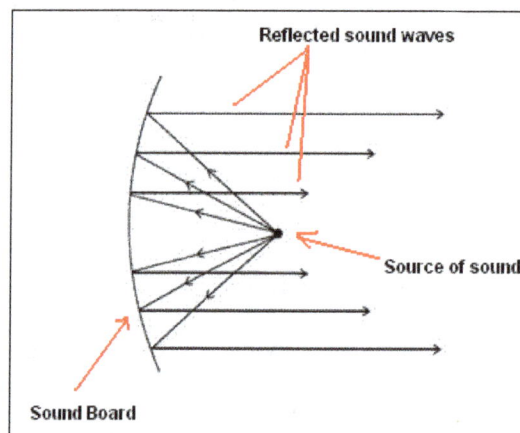

Reflected sound waves

Source of sound

Sound Board

Echo

The sound which we hear after reflection is called echo or echo of sound. One can hear the echo by shouting loudly in a big hall. After shouting loudly, the same sound reaches the ears after reflecting from the surface of the wall. Echo of sound can be heard by producing sound at place surrounded by hills or big buildings. Thus, repetition of sound because of multiple reflection of sound wave is called echo.

Condition Necessary for Creation of Echo

One does not experience any echo sound in a small room. This does not mean that sound does not get reflected in a small room, but necessary conditions for production of echo are not present. Any sound persists on one's brain upto 0.1 second of time. So echo can only be heard if the same sound comes to one's ear after a lapse of 0.1 second. Thus, reflection of sound must reach to the brain after a lapse of 0.1 second.

Explanation: Since, sound covers 344 m in air in 1 second.

Thus, in 0.1 second sound would cover a distance of 344 m x 0.1 = 34.4 m.

Thus, to hear an echo sound the reflecting surface must be at a distance of 17.2 m, so that sound has to cover a distance which is more than 17.2 m x 2 = 34.4 m; before reaching the ears.

So, if reflecting surface is at a distance of more than 17.2 m, the sound would reach to our brain after 0.1 second and we would be able to hear the echo of sound.

Thus, there are two conditions to experience the echo of sound.

- Sound must come back to the person after 0.1 second.

- For above condition, the reflecting surface must be at a minimum distance of 17.2m.

Multiple Echo

You may have heard the echo of your yahoo in hilly areas. This happens because of multiple reflection of sound wave and is often called multiple echoes.

The rolling sound of thunder is heard because of the multiple reflections of thunder sound or multiple echoes. The sound of thunder comes to us many times because of reflections from clouds and earth surface.

Use of Multiple Reflection of Sound

- In measuring the depth of sea/ocean.

- For the detection of the position of any objects, such as shipwrecks, sea rocks, hidden iceberg in the sea and ocean.

- Investigating any problem inside the human body.

For above mentioned purposes, sound of high frequency is produced so that reflections can be received from various surfaces. The time taken for reception of reflected sound waves is analyzed by a computer to detect the problem.

Acoustic Phenomenon

Acoustic waves are reflected by walls or other hard surfaces, such as mountains and privacy fences. The reason of reflection may be explained as a discontinuity in the propagation medium. This can be heard when the reflection returns with sufficient magnitude and delay to be perceived distinctly. When sound, or the echo itself, is reflected multiple times from multiple surfaces, the echo is characterized as a reverberation.

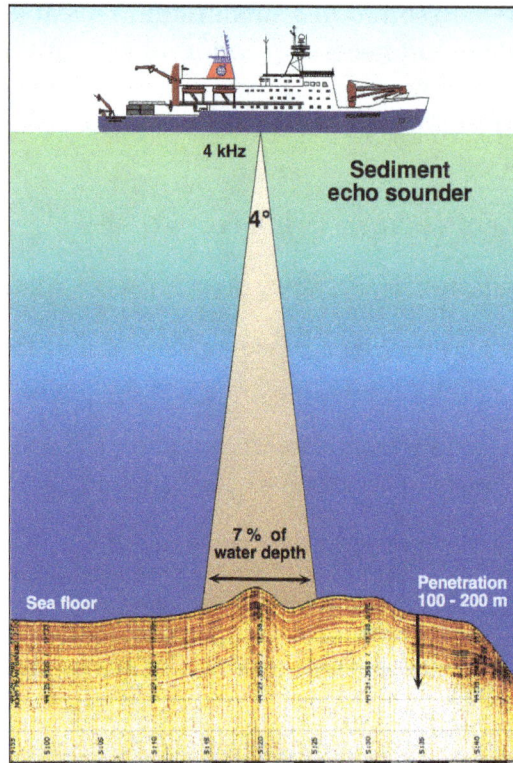

The principle of sediment echo sounding, which uses a
narrow beam of high energy and low frequency.

The human ear cannot distinguish echo from the original direct sound if the delay is less than 1/10 of a second. The velocity of sound in dry air is approximately 343 m/s at a temperature of 25 °C. Therefore, the reflecting object must be more than 17.2m from the sound source for echo to be perceived by a person located at the source. When a sound produces an echo in two seconds, the reflecting object is 343m away. In nature, canyon walls or rock cliffs facing water are the most common natural settings for hearing echoes. The strength of echo is frequently measured in dB sound pressure level (SPL) relative to the directly transmitted wave. Echoes may be desirable (as in sonar) or undesirable (as in telephone systems).

In Music

In music performance and recording, electric echo effects have been used since the 1950s. The Echoplex is a tape delay effect, first made in 1959 that recreates the sound of an acoustic echo. Designed by Mike Battle, the Echoplex set a standard for the effect in the 1960s and was used by most of the notable guitar players of the era; original Echoplexes are highly sought after. While Echoplexes were used heavily by guitar players (and the occasional bass player, such as Chuck Rainey, or trumpeter, such as Don Ellis), many recording studios also used the Echoplex. Beginning in the 1970s, Market built the solid-state Echoplex for Maestro. In the 2000s, most echo effects units use electronic or digital circuitry to recreate the echo effect.

Reverberation

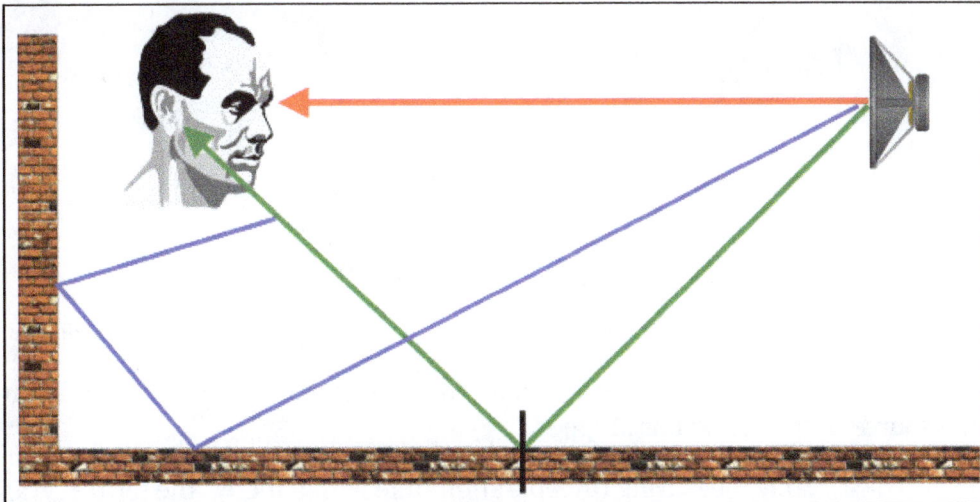

Reverberation, in psychoacoustics and acoustics, is a persistence of sound after the sound is produced. A reverberation, or reverb, is created when a sound or signal is reflected causing a large number of reflections to build up and then decay as the sound is absorbed by the surfaces of objects in the space – which could include furniture, people, and air. This is most noticeable when the sound source stops but the reflections continue, decreasing in amplitude, until they reach zero amplitude.

Reverberation is frequency dependent: The length of the decay, or reverberation time, receives special consideration in the architectural design of spaces which need to have specific reverberation times to achieve optimum performance for their intended activity. In comparison to a distinct echo, that is detectable at a minimum of 50 to 100 ms after the previous sound, reverberation is the occurrence of reflections that arrive in a sequence of less than approximately 50 ms. As time passes, the amplitude of the reflections gradually reduces to non-noticeable levels. Reverberation is not limited to indoor spaces as it exists in forests and other outdoor environments where reflection exists.

Reverberation occurs naturally when a person sings, talks, or plays an instrument acoustically in a hall or performance space with sound-reflective surfaces. The sound of reverberation is often electronically added to the vocals of singers and to musical instruments. This is done in both live sound systems and sound recordings by using effects units. Effects units that are specialized in the generation of the reverberation effect are commonly called reverbs.

Whereas reverberation normally adds to the naturalness of recorded sound by adding a sense of space, reverberation can reduce speech intelligibility, especially when noise is also present. Users of hearing aids frequently report difficult in understanding speech in reverberant, noisy situations. Reverberation is a very significant source of mistakes in automatic speech recognition. Dereverberation is the process of reducing the level of reverberation in a sound or signal.

Reverberation Time

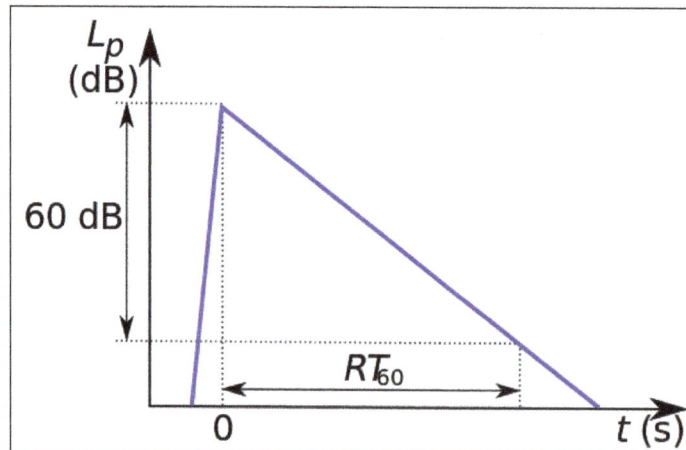

Sound level in a reverberant cavity excited by a pulse, as a function of time (very simplified diagram).

Reverberation time is a measure of the time required for the sound to "fade away" in an enclosed area after the source of the sound has stopped.

When it comes to accurately measuring reverberation time with a meter, the term T_{60} (an abbreviation for Reverberation Time 60dB) is used. T_{60} provides an objective reverberation time measurement. It is defined as the time it takes for the sound pressure level to reduce by 60 dB, measured after the generated test signal is abruptly ended.

Reverberation time is frequently stated as a single value if measured as a wideband signal (20 Hz to 20 kHz). However, being frequency dependent, it can be more precisely described in terms of frequency bands (one octave, 1/3 octave, 1/6 octave, etc). Being frequency dependent, the reverberation time measured in narrow bands will differ depending on the frequency band being measured. For precision, it is important to know what ranges of frequencies are being described by a reverberation time measurement.

In the late 19th century, Wallace Clement Sabine started experiments at Harvard University to investigate the impact of absorption on the reverberation time. Using a portable wind chest and organ pipes as a sound source, a stopwatch and his ears, he measured the time from interruption of the source to inaudibility (a difference of roughly 60 dB). He found that the reverberation time is proportional to room dimensions and inversely proportional to the amount of absorption present.

The optimum reverberation time for a space in which music is played depends on the type of music that is to be played in the space. Rooms used for speech typically need a shorter reverberation time so that speech can be understood more clearly. If the reflected sound from one syllable is still heard when the next syllable is spoken, it may be difficult to understand what was said. "Cat", "Cab", and "Cap" may all sound very similar. If on the other hand the reverberation time is too short, tonal balance and loudness may suffer. Reverberation effects are often used in studios to add depth to sounds. Reverberation changes the perceived spectral structure of a sound but does not alter the pitch.

Basic factors that affect a room's reverberation time include the size and shape of the enclosure as well as the materials used in the construction of the room. Every object placed within the enclosure can also affect this reverberation time, including people and their belongings.

Measurement of Reverberation Time

Historically, reverberation time could only be measured using a level recorder (a plotting device which graphs the noise level against time on a ribbon of moving paper). A loud noise is produced, and as the sound dies away the trace on the level recorder will show a distinct slope. Analysis of this slope reveals the measured reverberation time. Some modern digital sound level meters can carry out this analysis automatically.

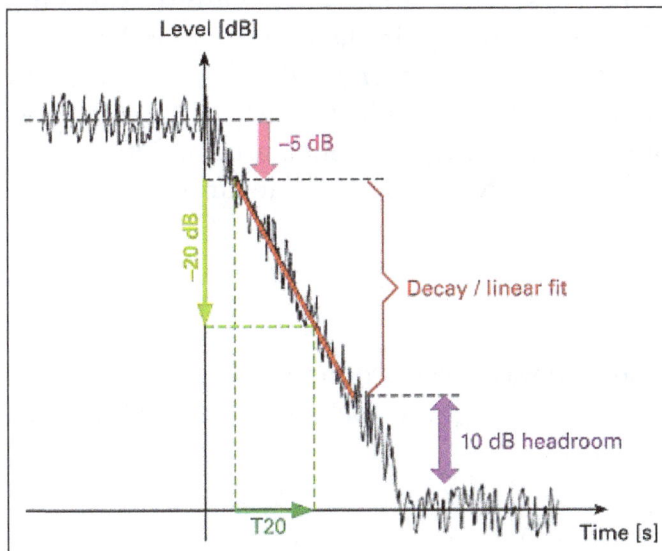

Automatically determining T20 value - 5dB trigger - 20dB measurement - 10dB headroom to noise floor.

Several methods exist for measuring reverb time. An impulse can be measured by creating a sufficiently loud noise (which must have a defined cut-off point). Impulse noise sources such as a blank pistol shot or balloon burst may be used to measure the impulse response of a room.

Alternatively, a random noise signal such as pink noise or white noise may be generated through a loudspeaker, and then turned off. This is known as the interrupted method, and the measured result is known as the interrupted response.

A two-port measurement system can also be used to measure noise introduced into a space and compare it to what is subsequently measured in the space. Consider sound reproduced by a loudspeaker into a room. A recording of the sound in the room can be made and compared to what was sent to the loudspeaker. The two signals can be compared mathematically. This two port measurement system utilizes a Fourier transform to mathematically derive the impulse response of the room. From the impulse response, the reverberation time can be calculated. Using a two-port system allows reverberation time to be measured with signals other than loud impulses. Music or recordings of other sounds can be used. This allows measurements to be taken in a room after the audience is present.

Reverberation time is usually stated as a decay time and is measured in seconds. There may or may not be any statement of the frequency band used in the measurement. Decay time is the time it takes the signal to diminish 60 dB below the original sound. It is often difficult to inject enough sound into the room to measure a decay of 60 dB, particularly at lower frequencies. If the decay is linear, it is sufficient to measure a drop of 20 dB and multiply the time by 3, or a drop of 30 dB and multiply the time by 2. These are the so-called T20 and T30 measurement methods.

The RT_{60} reverberation time measurement is defined in the ISO 3382-1 standard for performance spaces, the ISO 3382-2 standard for ordinary rooms, and the ISO 3382-3 for open-plan offices, as well as the ASTM E2235 standard.

The concept of Reverberation Time implicitly supposes that the decay rate of the sound is exponential, so that the sound level diminishes regularly, at a rate of so many dB per second. It is not often the case in real rooms, depending on the disposition of reflective, dispersive and absorbing surfaces. Moreover, successive measurement of the sound level often yields very different results, as differences in phase in the exciting sound build up in notably different sound waves. In 1965, Manfred R. Schroeder published "A new method of Measuring Reverberation Time" in the Journal of the Acoustical Society of America. He proposed to measure, not the power of the sound, but the energy, by integrating it. This made it possible to show the variation in the rate of decay and to free acousticians from the necessity of averaging many measurements.

Sabine Equation

Sabine's reverberation equation was developed in the late 1890s in an empirical fashion. He established a relationship between the T_{60} of a room, its volume, and its total absorption (in sabins). This is given by the equation:

$$T_{60} = \frac{24\ln 10^1}{c_{20}}\frac{V}{Sa} \approx 0.1611\,\text{sm}^{-1}\frac{V}{Sa},$$

where c_{20} is the speed of sound in the room (for 20 degrees Celsius), V is the volume of the room in m³, S total surface area of room in m², a is the average absorption coefficient of room surfaces, and the product Sa is the total absorption in sabins.

The total absorption in sabins (and hence reverberation time) generally changes depending on frequency (which is defined by the acoustic properties of the space). The equation does not take into account room shape or losses from the sound traveling through the air (important in larger spaces). Most rooms absorb less sound energy in the lower frequency ranges resulting in longer reverb times at lower frequencies.

Sabine concluded that the reverberation time depends upon the reflectivity of sound from various surfaces available inside the hall. If the reflection is coherent, the reverberation time of the hall will be longer; the sound will take more time to die out.

The reverberation time RT_{60} and the volume V of the room have great influence on the critical distance d_c (conditional equation):

$$d_c \approx 0.057 \cdot \sqrt{\frac{V}{RT_{60}}}$$

where critical distance d_c is measured in meters, volume V is measured in m³, and reverberation time RT_{60} is measured in seconds.

Absorption Coefficient

The absorption coefficient of a material is a number between 0 and 1 which indicates the proportion of sound which is absorbed by the surface compared to the proportion which is reflected back. the room. A large, fully open window would offer no reflection as any sound reaching it would pass straight out and no sound would be reflected. This would have an absorption coefficient of 1. Conversely, a thick, smooth painted concrete ceiling would be the acoustic equivalent of a mirror and have an absorption coefficient very close to 0.

Reverberation in Music Composition and Performance

Several composers employ the reverberation effect as a main sound resource, having a comparable relavance as the solo instrument. For example, Pauline Oliveros, Henrique Machado and many others. In order to employ the reverberant properties of the room, the composers are supposed to investigate and probe the sound response of that particular ambient, that will affect and inspire the creation of the musical work.

Creating Reverberation Effects

A performer or a producer of live or recorded music often induces reverberation in a work. Several systems have been developed to produce or to simulate reverberation.

Chamber Reverberators

The first reverb effects created for recordings used a real physical space as a natural echo chamber. A loudspeaker would play the sound, and then a microphone would pick it up again, including the effects of reverb. Although this is still a common technique, it requires a dedicated soundproofed room, and varying the reverb time is difficult.

Plate Reverberators

Transducer, similar to the driver in a loudspeaker, to create vibrations in a large plate of sheet metal. The plate's motion is picked up by one or more contact microphones whose output is an audio signal which may be added to the original "dry" signal. In the late 1950s, Elektro-Mess-Technik (EMT) introduced the EMT 140; a 600-pound (270 kg) model popular in recording studios, contributing to many hit records such as Beatles and Pink Floyd albums recorded at Abbey Road Studios in the 1960s, and others recorded by Bill Porter in Nashville's RCA Studio B. Early units had one pickup for mono output, and later models featured two pickups for stereo use. The reverb time can be adjusted by a damping pad, made from framed acoustic tiles. The closer the damping pad, the shorter the reverb time. However, the pad never touches the plate. Some units also featured a remote control.

Spring Reverberators

A spring reverb system uses a transducer at one end of a spring and a pickup at the other, similar to those used in plate reverbs, to create and capture vibrations within a metal spring. Laurens Hammond was granted a patent on a spring-based mechanical reverberation system in 1939. The Hammond Organ included a built-in spring reverberator.

Folded line reverberation device.

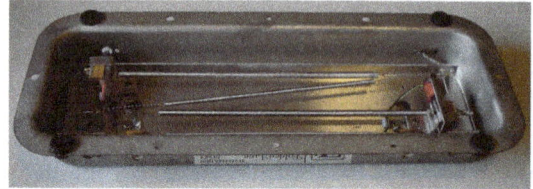

The folded coil spring is visible from the underside of the reverberation device.

Spring reverberators were once widely used in semi-professional recording and are frequently incorporated into Guitar amplifiers due to their modest cost and small size. One advantage over more sophisticated alternatives is that they lend themselves to the creation of special effects; for example rocking them back and forth creates a thundering, crashing sound caused by the springs colliding with each other.

Digital Reverberators

Digital reverberators use various signal processing algorithms in order to create the reverb effect. Since reverberation is essentially caused by a very large number of echoes, simple reverberation algorithms use several feedback delay circuits to create a large, decaying series of echoes. More advanced digital reverb generators can simulate the time and frequency domain response of a specific room (using room dimensions, absorption, and other properties). In a music hall, the direct sound always arrives at the listener's ear first because it follows the shortest path. Shortly after the direct sound, the reverberant sound arrives. The time between the two is called the "pre-delay."

Reverberation, or informally, "reverb" or "verb", is one of the most universally used audio effects and is often found in guitar pedals, synthesizers, effects units, digital audio workstations (DAWs) and VST plug-ins.

Convolution Reverb

Convolution reverb is a process used for digitally simulating reverberation. It uses the mathematical convolution operation, a pre-recorded audio sample of the impulse response of the space being modeled, and the sound to be echoed, to produce the effect. The impulse-response recording is first stored in a digital signal-processing system. This is then convolved with the incoming audio signal to be processed.

Applications of Multiple Reflections of Sound

The reflection of sound is utilized in the working of devices like megaphone, sound boards and ear trumpet:

A megaphone (or speaking-tube) is a horn-shaped tube, which is used to address a small gathering of people at places like tourist spots, fairs, and market places and during demonstrations. One end of the megaphone tube is narrow and its other end is quite wide. When a person speaks into the narrow end of the megaphone tube, the sound waves produced by his voice are prevented from spreading by successive reflections from the wider end of the megaphone tube. Due to this the sound of the voice of the person can be heard over a longer distance.

Thus, a megaphone (or speaking-tube) works on the reflection of sound.The loudspeakers also have horn-shaped opening so that the sound of the voice of speaker (or music) can be heard by a large gathering over a considerable distance.

The sound board is a concave board (curved board), which is placed behind the speaker in large halls or auditoriums so that his speech can be heard easily even by the persons sitting at a considerable distance. The sound board works as follows: The speaker is made to stand at the focus of the concave sound board. The concave surface of the sound board reflects the sound waves of the speaker towards the audience (and hence prevents the spreading of sound in various directions).

Due to this, sound is distributed uniformly throughout the hall and even the persons sitting at the back of the hall can hear his speech easily. It is obvious that the sound boards work on the reflection of sound.

The ear trumpet is a hearing aid, which is used by the persons who are hard of hearing. One end of the ear trumpet is wide, whereas its other end is narrow. The sound waves from a large surrounding area fall on the wide end of the trumpet, which reflects them into its narrower end that leads into the ear. In this way, more sound energy falls on the eardrum of the person and that leads into the ear. In this way, more sound energy falls on the eardrum of the person and improves his hearing capacity. Thus, an ear trumpet also works on the reflection of sound.

In a stethoscope the heart beat of a patient reaches the doctor's ears by multiple reflection of sound.

Refraction of Sound

Suppose you are camping on the shore of a lake which is not too wide, maybe 1/2 a mile across or so. During the day you can see campers on the other side of the lake, but you cannot hear them. At night, however, you can not only see the campers on the other side of the lake but you can also hear their conversations as they sit around their camp fire. This phenomena is due to the refraction of sound waves.

The speed of a wave depends on the elastic and inertia properties of the medium through which it travels. When a wave encounters different medium where the wave speed is different, the wave will change directions. Most often refraction is encountered in a study of optics, with a ray of light incident upon a boundary between two media (air and glass, or air and water, or glass and water). Snell's law relates the directions of the wave before and after it crosses the boundary between the two media.

$$\frac{\sin\theta_1}{c_1} = \frac{\sin\theta_2}{c_2}$$

Notice that as the wavefronts cross the boundary the wavelength changes, but the frequency remains constant.

In acoustics, however, sound waves usually don't encounter an abrupt change in medium properties. Instead the wave speed changes gradually over a given distance. The speed of a sound wave in air depends on the temperature (c=331 + 0.6 T) where T is the temperature in °C. Often the change in the wave speed, and the resulting refraction, is due to a change in the local temperature of the air. For example, during the day the air is warmest right next to the ground and grows cooler above the ground. This is called a temperature lapse. Since the temperature decreases with height, the speed of sound also decreases with height. This means that for a sound wave traveling close to the ground, the part of the wave closest to the ground is traveling the fastest, and the part of the wave farthest above the ground is traveling the slowest. As a result, the wave changes direction and bends upwards. This can create a "shadow zone" region into which the sound wave cannot penetrate. A person standing in the shadow zone will not hear the sound even though he/she might be able to see the source. The sound waves are being refracted upwards and will never reach the observer.

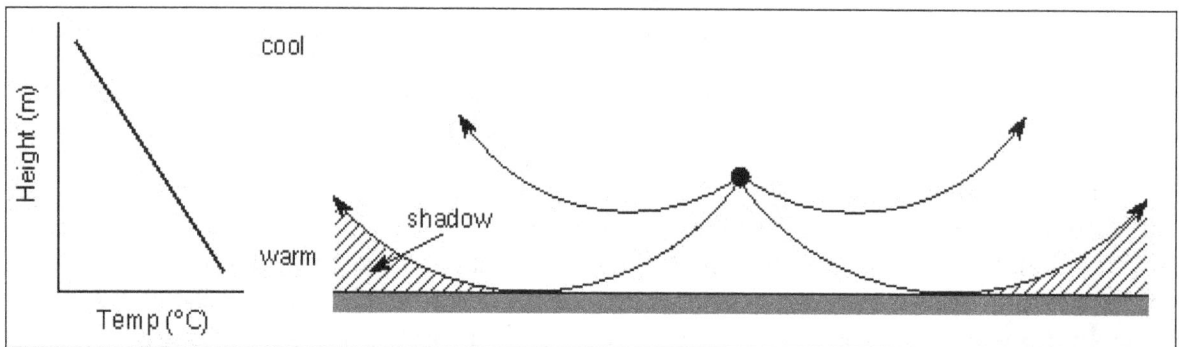

A temperature inversion is when the temperature is coolest right next to the ground and warmer as you increase in height above the ground. Since the temperature increases with height, the speed of sound also increases with height. This means that for a sound wave traveling close to the ground, the part of the wave closest to the ground is traveling the slowest, and the part of the wave farthest above the ground is traveling the fastest. As a result, the wave changes direction and bends downwards. Temperature inversions most often happen at night after the sun goes down when the ground (or water in a lake) cools off quickly, while the air above the ground remains warm. This downward refraction of sound is why you can hear the conversations of campers across the lake, when otherwise you should not be able to hear them.

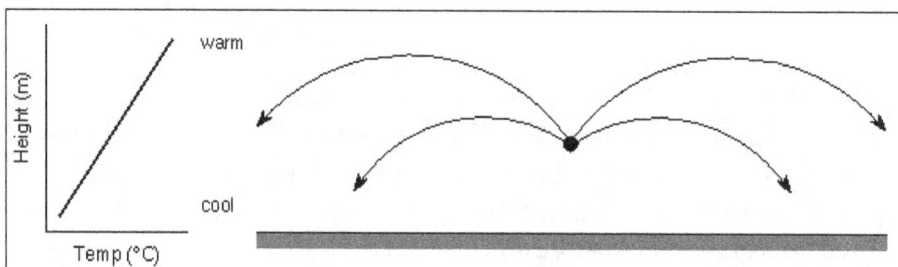

Diffraction of Sound

Diffraction is the bending of waves around small* obstacles and the spreading out of waves beyond small openings.

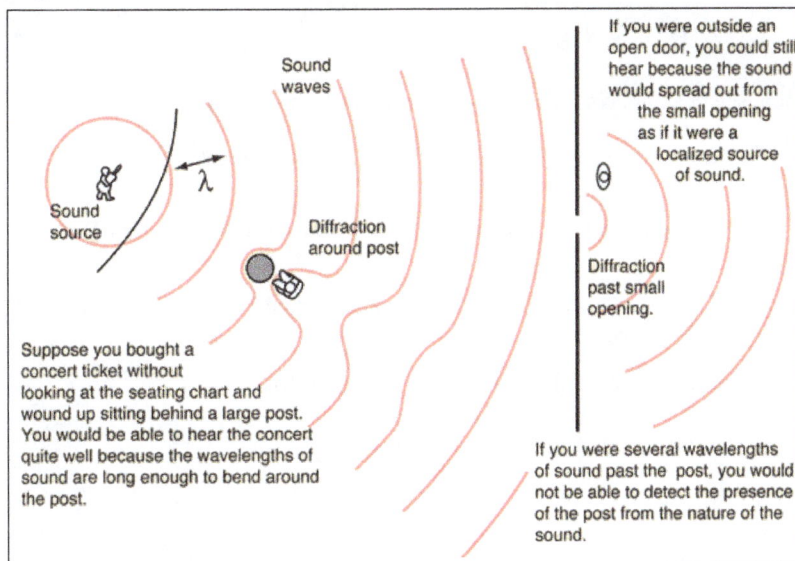

Important parts of our experience with sound involve diffraction. The fact that you can hear sounds around corners and around barriers involves both diffraction and reflection of sound. Diffraction in such cases helps the sound to "bend around" the obstacles. The fact that diffraction is more pronounced with longer wavelengths implies that you can hear low frequencies around obstacles better than high frequencies, as illustrated by the example of a marching band on the street. Another common example of diffraction is the contrast in sound from a close lightning strike and a distant one. The thunder from a close bolt of lightning will be experienced as a sharp crack, indicating the presence of a lot of high frequency sound. The thunder from a distant strike will be experienced as a low rumble since it is the long wavelengths which can bend around obstacles to get to you. There are other factors such as the higher air absorption of high frequencies involved, but diffraction plays a part in the experience.

You may perceive diffraction to have a dual nature, since the same phenomenon which causes waves to bend around obstacles causes them to spread out past small openings. This aspect of diffraction also has many implications. Besides being able to hear the sound when you are outside the door as in the illustration above, this spreading out of sound waves has consequences when you are trying to soundproof a room. Good soundproofing requires that a room be well sealed, because any openings will allow sound from the outside to spread out in the room - it is surprising how much sound can get in through a small opening. Good sealing of loudspeaker cabinets is required for similar reasons.

Another implication of diffraction is the fact that a wave which is much longer than the size of an obstacle, like the post in the auditorium above, cannot give you information about that obstacle. A fundamental principle of imaging is that you cannot see an object which is smaller than the wavelength of the wave with which you view it. You cannot see a virus with a light microscope because the virus is smaller than the wavelength of visible light. The reason for that limitation can be visualized with the auditorium example: the sound waves bend in and reconstruct the wavefront past the post. When you are several sound wavelengths past the post, nothing about the wave gives you information about the post. So your experience with sound can give you insights into the limitations of all kinds of imaging processes.

References

- Soundreflection, sciencenine, classnine: excellup.com, Retrieved 21 July, 2019

- Wölfel, Matthias; mcdonough, John (2009). Distant Speech Recognition. Chichester: John Wiley & Sons. P. 48. ISBN 0470714077

- So why does reverberation affect speech intelligibility?". MC Squared System Design Group, Inc. Retrieved 2008-12-04

- Davis, Gary (1987). The sound reinforcement handbook (2nd ed.). Milwaukee, WI: Hal Leonard. P. 259. ISBN 9780881889000. Retrieved February 12, 2016

- Refract, Demos, drussell: acs.psu.edu, Retrieved 15 April, 2019

8

Acoustic Transmission

The transmission of sounds through and between materials, such as walls, air and musical instruments, is referred to as acoustic transmission. Some of the major concepts related to acoustic transmission are sound transmission class and sound transmission loss. This chapter has been carefully written to provide an easy understanding of these facets of acoustic transmission.

Sound waves are pressure waves that travel through Earth's crust, water bodies, and atmosphere. Natural sound frequencies specify the frequency attributes of sound waves that will efficiently induce vibration in a body (e.g., the tympanic membrane of the ear) or that naturally result from the vibration of that body.

Sound waves are created by a disturbance that then propagates through a medium (e.g., crust, water, air). Individual particles are not transmitted with the wave, but the propagation of the wave causes particles (e.g., individual air molecules) to oscillate about an equilibrium position.

Every object has a unique natural frequency of vibration. Vibration can be induced by the direct forcible disturbance of an object or by the forcible disturbance of the medium in contact with an object (e.g. the surrounding air or water). Once excited, all such vibrators (i.e., vibratory bodies) become generators of sound waves. For example, when a rock falls, the surrounding air and impacted crust undergo sinusoidal oscillations and generate a sound wave.

Vibratory bodies can also absorb sound waves. Vibrating bodies can, however, efficiently vibrate only at certain frequencies called the natural frequencies of oscillation. In the case of a tuning fork, if a traveling sinusoidal sound wave has the same frequency as the sound wave naturally produced by the oscillations of the tuning fork, the traveling pressure wave can induce vibration of the tuning fork at that particular frequency.

Mechanical resonance occurs with the application of a periodic force at the same frequency as the natural vibration frequency. Accordingly, as the pressure fluctuations in a resonant traveling sound wave strike the prongs of the fork, the prongs experience successive forces at appropriate intervals to produce sound generation at the natural vibrational or natural sound frequency. If the resonant traveling wave continues to exert force, the amplitude of oscillation of the tuning fork will increase and the sound wave emanating from the tuning fork will grow stronger. If the frequencies are within the range of human hearing, the sound will seem to grow louder. Singers are able to break glass by loudly singing a note at the natural vibrational frequency of the glass. Vibrations induced in the glass can become so strong that the glass exceeds its elastic limit and breaks. Similar phenomena occur in rock formations.

All objects have a natural frequency or set of frequencies at which they vibrate.

Sound waves can potentiate or cancel in accord with the principle of superposition and whether they are in phase or out of phase with each other. Waves of all forms can undergo constructive or destructive interference. Sound waves also exhibit Doppler shifts—an apparent change in frequency due to relative motion between the source of sound emission and the receiving point. When sound waves move toward an observer the Doppler effect shifts observed frequencies higher. When sound waves move away from an observer the Doppler effect shifted observed frequencies lower. The Doppler effect is commonly and easily observed in the passage of planes, trains, and automobiles.

The speed of propagation of a sound wave is dependent upon the density of the medium of transmission. Weather conditions (e.g., temperature , pressure, humidity, etc.) and certain geophysical and topographical features (e.g., mountains or hills) can obstruct sound transmission. The alteration of sound waves by commonly encountered meteorological conditions is generally negligible except when the sound waves propagate over long distances or emanate from a high frequency source. In the extreme cases, atmospheric conditions can bend or alter sound wave transmission.

The speed of sound through a fluid—inclusive in this definition of "fluid" are atmospheric gases—depends upon the temperature and density of the fluid. Sound waves travel faster at higher temperature and density of medium. As a result, in a standard atmosphere, the speed of sound (reflected in the Mach number) lowers with increasing altitude.

Meteorological conditions that create layers of air at dramatically different temperatures can refract sound waves.

The speed of sound in water is approximately four times faster than the speed of sound in air. SONAR sounding of ocean terrain is a common tool of oceanographers. Properties such as pressure, temperature, and salinity also affect the speed of sound in water.

Because sound travels so well under water, many marine biologists argue that the introduction of man-made noise (e.g., engine noise, propeller cavitation, etc) into the oceans within the last two centuries interferes with previously evolutionarily well-adapted methods of sound communication between marine animals. For example, man-made noise has been demonstrated to interfere with long-range communications of whales. Although the long term implications of this interference are not fully understood, many marine biologists fear that this interference could impact whale mating and lead to further population reductions or extinction.

Acoustic Transmission Line

An acoustic transmission line is the use of a long duct, which acts as an acoustic waveguide and is used to produce or transmit sound in an undistorted manner. Technically it is the acoustic analog of the electrical transmission line, typically conceived as a rigid-walled duct or tube, that is long and thin relative to the wavelength of sound present in it.

Examples of transmission line (TL) related technologies include the (mostly obsolete) speaking tube, which transmitted sound to a different location with minimal loss and distortion, wind instruments such as the pipe organ, woodwind and brass which can be modeled in part as transmission lines (although their design also involves generating sound, controlling its timbre, and coupling it efficiently to the open air), and transmission line based loudspeakers which use the same principle to produce accurate extended low bass frequencies and avoid distortion. The comparison between an acoustic duct and an electrical transmission line is useful in "lumped-element" modeling of acoustical systems, in which acoustic elements like volumes, tubes, pistons, and screens can be modeled as single elements in a circuit. With the substitution of pressure for voltage, and volume particle velocity for current, the equations are essentially the same. Electrical transmission lines can be used to describe acoustic tubes and ducts, provided the frequency of the waves in the tube is below the critical frequency, such that they are purely planar.

Design Principles

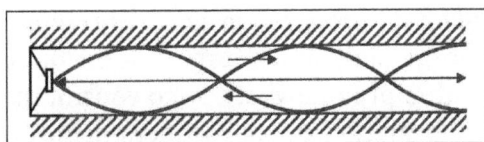

Relationship between TL length and wavelength

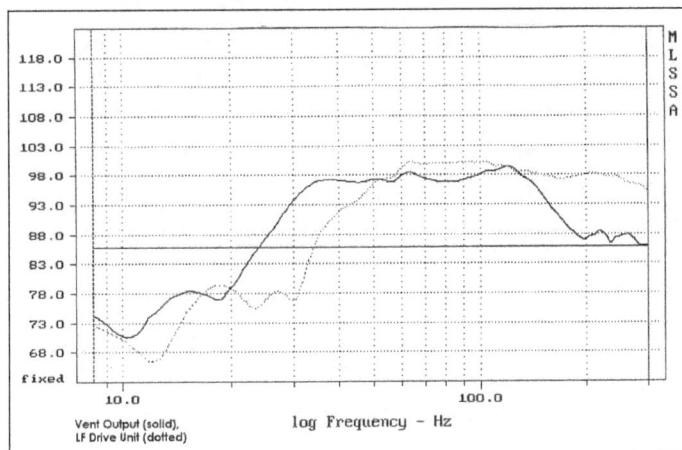

Frequency response (magnitude) measurement of drive unit and TL outputs

Phase inversion is achieved by selecting a length of line that is equal to the quarter wavelength of the target lowest frequency. Which shows a hard boundary at one end (the speaker) and the open-ended line vent at the other. The phase relationship between the bass driver and vent is in phase in the pass band until the frequency approaches the quarter wavelength, when the relationship reaches 90 degrees as shown. However by this time the vent is producing most of the output. Because the line is operating over several octaves with the drive unit, cone excursion is reduced, providing higher SPL's and lower distortion levels, compared with reflex and infinite baffle designs.

The calculation of the length of the line required for a certain bass extension appears to be straightforward, based on a simple formula:

$$\ell = \frac{344}{4 \times f},$$

where:

f is the sound frequency in Hertz (Hz)

344 $^m/_s$ is the speed of sound in air at 20 °C

ℓ is the length of the transmission line in meters.

The complex loading of the bass drive unit demands specific Thiele-Small driver parameters to realise the full benefits of a TL design. Most drive units in the marketplace are developed for the more common reflex and infinite baffle designs and are usually not suitable for TL loading. High efficiency bass drivers with extended low frequency ability, are usually designed to be extremely light and flexible, having very compliant suspensions. Whilst performing well in a reflex design, these characteristics do not match the demands of a TL design. The drive unit is effectively coupled to a long column of air which has mass. This lowers the resonant frequency of the drive unit, negating the need for a highly compliant device. Furthermore, the column of air provides greater force on the driver itself than a driver opening onto a large volume of air (in simple terms it provides more resistance to the driver's attempt to move it), so to control the movement of air requires an extremely rigid cone, to avoid deformation and consequent distortion.

The introduction of the absorption materials reduces the velocity of sound through the line, as discovered by Bailey in his original work. Bradbury published his extensive tests to determine this effect in a paper in the Journal of the Audio Engineering Society (JAES) in 1976 and his results agreed that heavily damped lines could reduce the velocity of sound by as much as 50%, although 35% is typical in medium damped lines. Bradbury's tests were carried out using fibrous materials, typically longhaired wool and glass fibre. These kinds of materials, however, produce highly variable effects that are not consistently repeatable for production purposes. They are also liable to produce inconsistencies due to movement, climatic factors and effects over time. High-specification acoustic foams, developed by loudspeaker manufacturers such as PMC, with similar characteristics to longhaired wool, provide repeatable results for consistent production. The density of the polymer, the diameter of the pores and the sculptured profiling are all specified to provide the correct absorption for each speaker model. Quantity and position of the foam is critical to engineer a low-pass acoustic filter that provides adequate attenuation of the upper bass frequencies, whilst allowing an unimpeded path for the low bass frequencies.

Uses

Loudspeaker Design

Acoustic transmission lines gained attention in their use within loudspeakers in the 1960s and 1970s. In 1965, A R Bailey's article in Wireless World, "A Non-resonant Loudspeaker Enclosure Design", detailed a working Transmission Line, which was commercialized by John Wright and partners under the brand name IMF and later TDL, and were sold by audiophile Irving M. "Bud" Fried in the United States.

A transmission line is used in loudspeaker design, to reduce time, phase and resonance related distortions, and in many designs to gain exceptional bass extension to the lower end of human hearing, and in some cases the near-infrasonic (below 20 Hz). TDL's 1980s reference speaker range (now discontinued) contained models with frequency ranges of 20 Hz upwards, down to 7 Hz upwards, without needing a separate subwoofer. Irving M. Fried, an advocate of TL design, stated that:

> "I believe that speakers should preserve the integrity of the signal waveform and the Audio Perfectionist Journal has presented a great deal of information about the importance of time domain performance in loudspeakers. I'm not the only one who appreciates time- and phase-accurate speakers but I have been virtually the only advocate to speak out in print in recent years. There's a reason for that."

In practice, the duct is folded inside a conventional shaped cabinet, so that the open end of the duct appears as a vent on the speaker cabinet. There are many ways in which the duct can be folded and the line is often tapered in cross section to avoid parallel internal surfaces that encourage standing waves. Depending upon the drive unit and quantity – and various physical properties – of absorbent material, the amount of taper will be adjusted during the design process to tune the duct to remove irregularities in its response. The internal partitioning provides substantial bracing for the entire structure, reducing cabinet flexing and colouration. The inside faces of the duct or line, are treated with an absorbent material to provide the correct termination with frequency to load the drive unit as a TL. A theoretically perfect TL would absorb all frequencies entering the line from the rear of the drive unit but remains theoretical, as it would have to be infinitely long. The physical constraints of the real world, demand that the length of the line must often be less than 4 meters before the cabinet becomes too large for any practical applications, so not all the rear energy can be absorbed by the line. In a realized TL, only the upper bass is TL loaded in the true sense of the term (i.e. fully absorbed); the low bass is allowed to freely radiate from the vent in the cabinet. The line therefore effectively works as a low-pass filter, another crossover point in fact, achieved acoustically by the line and its absorbent filling. Below this "crossover point" the low bass is loaded by the column of air formed by the length of the line. The length is specified to reverse the phase of the rear output of the drive unit as it exits the vent. This energy combines with the output of the bass unit, extending its response and effectively creating a second driver.

Sound Ducts as Transmission Lines

A duct for sound propagation also behaves like a transmission line (e.g. air conditioning duct, car muffler). Its length may be similar to the wavelength of the sound passing through it, but the dimensions of its cross-section are normally smaller than one quarter the wavelength. Sound is introduced at one end of the tube by forcing the pressure across the whole cross-section to vary with time. An almost planar wavefront travels down the line at the speed of sound. When the wave reaches the end of the transmission line, behaviour depends on what is present at the end of the line. There are three possible scenarios:

1. The frequency of the pulse generated at the transducer results in a pressure peak at the terminus exit (odd ordered harmonic open pipe resonance) resulting in effectively low acoustic impedance of the duct and high level of energy transfer.

2. The frequency of the pulse generated at the transducer results in a pressure null at the

terminus exit (even ordered harmonic open pipe anti -resonance) resulting in effectively high acoustic impedance of the duct and low level of energy transfer.

3. The frequency of the pulse generated at the transducer results in neither a peak or null in which energy transfer is nominal or in keeping with typical energy dissipation with distance from the source.

Sound Transmission Class

Sound Transmission Class (or STC) is an integer rating of how well a building partition attenuates airborne sound. In the US, it is widely used to rate interior partitions, ceilings, floors, doors, windows and exterior wall configurations. Outside the US, the Sound Reduction Index (SRI) ISO index is used. The STC rating figure very roughly reflects the decibel reduction in noise that a partition can provide. To improve the sound transmission class and increase soundproofing two techniques are employed; sound isolation and sound absorption. Sound isolation entails blocking the path that sound can travel through by using cavity walls, adding resilient channels or using caulking to seal a room. Sound absorption entails adding extra mass to an already isolated partition, often using extra gypsum layers or by placing rockwool into cavities.

The STC or sound transmission class is a single number method of rating how well wall partitions reduce sound transmission. The STC provides a standardized way to compare products such as doors and windows made by competing manufacturers. A higher number indicates better soundproofing than a lower number. The STC is a standardised theoretical measurement provided by ASTM E413 and E90 with the Field sound transmission class provided by ASTM E336-97 annex a1.

Sound Proofing

STC is used as a means of measuring the sound proofing of partitions between living spaces.

STC	What can be heard
25	Normal speech can be understood
30	Loud speech can be understood
35	Loud speech audible but not intelligible
40	Loud speech audible as a murmur
45	Loud speech heard but not audible
50	Loud sounds faintly heard
60+	Good soundproofing; most sounds do not disturb neighbouring residents.

Rating Methodology

The STC number is derived from sound attenuation values tested at sixteen standard frequencies from 125 Hz to 4000 Hz. These Transmission Loss values are then plotted on a sound pressure level graph and the resulting curve is compared to a standard reference contour provided by the ASTM.

Sound isolation metrics, such as the STC, are measured in specially-isolated and designed laboratory test chambers. There are nearly infinite field conditions that will affect sound isolation on site when designing building partitions and enclosures.

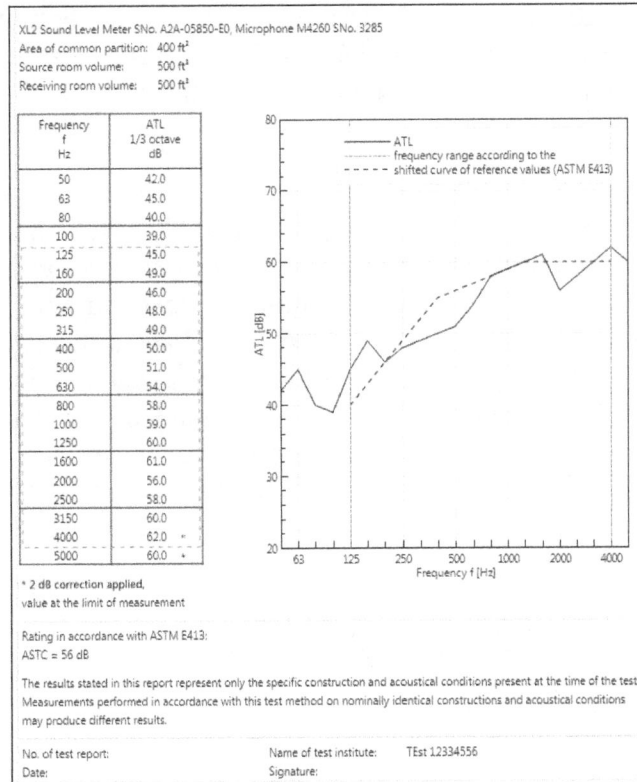

Sound Transmission Class Report Sample from *NTi Audio* showing
Transmission Loss in the sixteen standard frequencies

Factors Affecting Sound Transmission Class

The sound transmission class and thereby soundproofing is affected by the two factors; sound isolation and sound absorption.

Sound Isolation

Sound travels from one area to another in two ways. It travels through the air and it travels mechanically through the mass of the structure. To eliminate air borne sound all air paths between the areas must be eliminated. This is achieved by making seams airtight and closing all sound leaks. To eliminate structure-borne noise one must create isolation systems that reduce mechanical connections between those structures.

Structurally decoupling the gypsum wallboard panels from the partition framing can result in a large increase in sound isolation when installed correctly. Examples of structural decoupling in building construction include resilient channels, sound isolation clips and hat channels, and staggered- or double-stud framing. The STC results of decoupling in wall and ceiling assemblies varies significantly depending on the framing type, air cavity volume, and decoupling material type. Great care must be taken in each type of decoupled partition construction, as any fastener that becomes mechanically (rigidly) coupled to the framing can short-circuit the decoupling and result in drastically lower sound isolation results.

Sound damping tapes and caulking have been used to improve sound isolation since the early 1930s. Although the applications of sound damping tapes was largely limited to defense and industrial applications such as naval vessels and aircraft in the past, recent research has proven the effectiveness of damping in interior sound isolation in buildings.

Sound Leakage

For sound isolation to be effective all holes and gaps should be filled and the enclosure hermetically sealed. The table below illustrates sound proofing test results from a wall partition that has a theoretical maximum loss of 40dB from one room to the next and a partition area of 10 metres squared. Even small open gaps and holes in the partition have a disproportionate reduction in sound proofing. With 5% or 0.5 metres squared of the partition "open" and offering unrestricted sound transmission from one room to the next transmission loss reduces from 40dB to 13dB. A 0.1% open area, or 1cm squared, where for example caulking has not been applied effectively will reduce the transmission loss from 40dB to 30dB. Partitions that are inadequately sealed and contain back-to-back electrical boxes, untreated recessed lighting and unsealed pipes offer flanking paths for sound and significant leakage.

Transmission loss	% of area open
13dB loss	5% open
17dB loss	2% open
20dB loss	1% open
23db loss	0.5% open
27 dB loss	0.2% open
30dB loss	0.1% open
33dB loss	0.05% open
37dB loss	0.02% open
39.5dB loss	Practical maximum loss
40dB loss	Theoretical maximum loss

Sound Absorption

Sound absorption entails turning acoustical energy into some other form of energy, usually heat.

Mass

Adding mass to a partition reduces the transmission of sound. This is often achieved by adding additional layers of gypsum. The effect of adding multiple layers of gypsum wallboard to a frame also varies depending on the framing type and configuration. Doubling the mass of a partition does not double the STC, as the STC is calculated from a non-linear decibel sound transmission loss measurement. So, whereas installing an additional layer of gypsum wallboard to a light-gauge (25-ga. or lighter) steel stud partition will result in about a 5 STC-point increase, doing the same on single wood or single heavy-gauge steel will result in only 2 to 3 additional STC points. Adding a second additional layer (to the already 3-layer system) does not result in as drastic an STC change as the first additional layer. The effect of additional gypsum wallboard layers on double- and staggered-stud partitions is similar to that of light-gauge steel partitions. Due to increased mass, poured concrete and concrete blocks typically achieve higher STC values (in the mid STC 40s to the mid STC 50s) than equally thick framed walls. However the additional weight, added complexity of construction, and poor thermal insulation tend to limit masonry wall partitions as a viable sound isolation solution in many building construction projects.

Damping

Adding absorptive materials to the interior surfaces of rooms, for example fabric-faced fiberglass panels and thick curtains, will result in a decrease of reverberated sound energy within the room. However, absorptive interior surface treatments of this kind do not significantly improve the sound transmission class. Installing absorptive insulation, for example fiberglass batts and blow-in cellulose, into the wall or ceiling cavities does increase the sound transmission class significantly. The presence of insulation in single 2x4 wood stud framing spaced 16" (406 mm) on-center results in only a few STC points. In contrast, adding standard fiberglass insulation to an otherwise empty cavity in light-gauge (25-gauge or lighter) steel stud partitions can result in a nearly 10 STC-point improvement. As the stud gauge becomes heavier, the presence and type of insulation matters less.

Common Partition STC

Interior walls with 1 sheet of 1/2" (13 mm) gypsum wallboard (drywall) on either side of 2x4 (90 mm) wood studs spaced 16" (406 mm) on-center with fiberglass insulation filling each stud cavity have an STC of about 33. When asked to rate their acoustical performance, people often describe these walls as "paper thin." They offer little in the way of privacy. Double stud partition walls are typically constructed with varying gypsum wallboard panel layers attached to both sides of double 2x4 (90 mm) wood studs spaced 16" (406 mm) on-center and separated by a 1" (25 mm) airspace. These walls vary in sound isolation performance from the mid STC-40s into the high STC-60s depending on the presence of insulation and the gypsum wallboard type and quantity. Commercial buildings are typically constructed using steel studs of varying widths, gauges, and on-center spacings. Each of these framing characteristics have an effect on the sound isolation of the partition to varying degrees.

STC	Partition type
27	Single pane glass window (typical value) (Dual pane glass window range is 26-32)"*STC Ratings*".
33	Single layer of 1/2" drywall on each side, wood studs, no insulation (typical interior wall).
39	Single layer of 1/2" drywall on each side, wood studs, fiberglass insulation.
44	4" Hollow CMU (Concrete Masonry Unit).
45	Double layer of 1/2" drywall on each side, wood studs, batt insulation in wall.
46	Single layer of 1/2" drywall, glued to 6" lightweight concrete block wall, painted both sides.

46	6" Hollow CMU (Concrete Masonry Unit).
48	8" Hollow CMU (Concrete Masonry Unit).
50	10" Hollow CMU (Concrete Masonry Unit).
52	8" Hollow CMU (Concrete Masonry Unit) with 2" Z-Bars and 1/2" Drywall on each side.
54	Single layer of 1/2" drywall, glued to 8" dense concrete block wall, painted both sides.
54	8" Hollow CMU (Concrete Masonry Unit) with 1 1/2" Wood Furring, 1 1/2" Fiberglass Insulation and 1/2" Drywall on each side.
55	Double layer of 1/2" drywall on each side, on staggered wood stud wall, batt insulation in wall.
59	Double layer of 1/2" drywall on each side, on wood stud wall, resilient channels on one side, batt insulation.
63	Double layer of 1/2" drywall on each side, on double wood/metal stud walls (spaced 1" apart), double batt insulation.
64	8" Hollow CMU (Concrete Masonry Unit) with 3" Steel Studs, Fiberglass Insulation and 1/2" Drywall on each side.
72	8" concrete block wall, painted, with 1/2" drywall on independent steel stud walls, each side, insulation in cavities.

Sound Proofed Partitions

- Single metal stud partitions are more effective than single wood stud partitions, achieving a gain of between 2 and 10 STC. However for double stud partitions there is little difference between the two.

- A fibre or rock wool cavity fill will increase STC by between 5 and 8.

- Resilient channels are more effective on wood than metal studs.

- Increasing partition stud offset from 16 to 24 inches increases STC by 2 to 3 points.

- It is preferable to have non symmetrical leaves, for example with different thickness gypsum.

- Double stud partitions have a higher STC than single stud.

Transmission Loss

Transmission loss (TL) in duct acoustics, together with insertion loss (IL), describes the acoustic performances of a muffler-like system. It is frequently used in the industry areas such as muffler manufacturers and NVH (noise, vibration and harshness) department of automobile manufacturers. Generally the higher transmission loss of a system it has, the better it will perform in terms of noise cancellation.

Transmission loss (TL) (more specifically in duct acoustics) is defined as the difference between the power incident on a duct acoustic device (muffler) and that transmitted downstream into an anechoic termination. Transmission loss is independent of the source and presumes (or requires) an anechoic termination at the downstream end.

Transmission loss does not involve the source impedance and the radiation impedance inasmuch as it represents the difference between incident acoustic energy and that transmitted into an anechoic environment. Being made independent of the terminations, TL finds favor with researchers who are sometimes interested in finding the acoustic transmission behavior of an element or a set of elements in isolation of the terminations. But measurement of the incident wave in a standing wave acoustic field requires uses of impedance tube technology, may be quite laborious, unless one makes use of the two-microphone method with modern instrumentation.

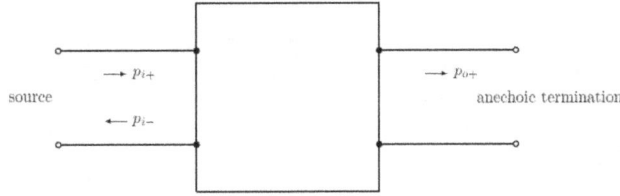

Transmission loss (duct acoustics) definition illustration.

By definition the TL on an acoustic component, for example a muffler, is described as:

$$TL = L_{Wi} - L_{Wo} = 10\log_{10}\left|\frac{S_i p_{i+} v_{i+}}{2}\frac{2}{S_o p_o v_0}\right| = 10\log_{10}\left|\frac{S_i p_{i+}^2}{S_o p_o^2}\right|$$

where:

- L_{Wi} is the incident sound power in the inlet coming towards muffler;

- L_{Wo} is the transmitted sound power going downstream in the outlet out of the muffler;

- S_i, S_o stand for the cross-sectional area of the inlet and outlet of muffler;

- p_{i+} is the acoustic pressure of the incident wave in the inlet, towards muffler;

- p_o is the acoustic pressure of the transmitted wave in the outlet, away from muffler.

- v_{i+} is the particle velocity of the incident wave in the inlet, towards muffler;

- v_o is the particle velocity of the transmitted wave in the outlet, away from muffler.

Note that p_{i+} cannot be measured directly in isolation from the reflected wave pressure p_{i-} (in the inlet, away from muffler). One has to resort to impedance tube technology or two-microphone method with modern instrumentation. However at the downstream side of the muffler, $p_o = p_{o+}$ in view of the anechoic termination, which ensures $p_{o-} = 0$.

And in most muffler applications, Si and So, the area of the exhaust pipe and tail pipe, are generally made equal, thus we have:

$$TL = 20\log_{10}\left|\frac{p_{i+}}{p_o}\right|$$

Thus, TL equals 20 times the logarithm (to the base 10) of the ratio of the acoustic pressure associated with the incident wave (in the exhaust pipe) and that of the transmitted wave (in the tail pipe), with the two pipes having the same cross-sectional area and the tail pipe terminating anechoically. However this anechoic condition is normally difficult to meet under practical industry environment, thus it is usually more convenient for the muffler manufacturers to measure insertions loss during their muffler performance tests under working conditions (mounted on an engine).

Also, since the transmitted sound power cannot possibly exceed the incident sound power (or $|p_{i+}|$ is always larger than $|p_o|$), it is known that TL will never be less than 0 dB.

Transmission Matrix

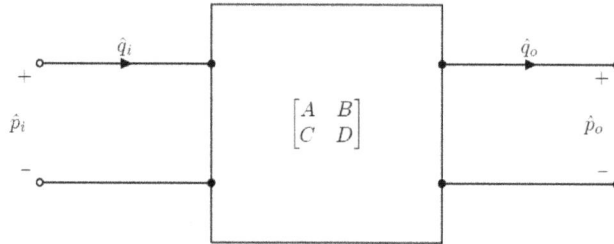

Transmission loss (duct acoustics) definition illustration with transmission matrix.

The low-frequency approximation implies that each subsystem is an acoustic two-port (or four-pole system) with two (and only two) unknown parameters, the complex amplitudes of two interfering waves travelling in opposite directions. Such a system can be described by its transmission matrix (or four-pole matrix), as follows:

$$\begin{bmatrix} \hat{p}_i \\ \hat{q}_i \end{bmatrix} = \begin{bmatrix} A & B \\ C & D \end{bmatrix} \begin{bmatrix} \hat{p}_o \\ \hat{q}_o \end{bmatrix},$$

where $\hat{p}_i$, $\hat{p}_i$, $\hat{p}_i$ and $\hat{q}_o$ are the sound pressures and volume velocities at the input and at the output. A, B, C and D are complex numbers. With this representation it can be prove that the transmission loss (TL) of this subsystem can be calculated as,

$$TL = 10 \log_{10} \left(\frac{1}{4} \left| A + B\frac{S}{\rho c} + C\frac{\rho c}{S} + D \right|^2 \right),$$

where:

- S is inlet and outlet cross-sectional area;
- ρc are media density and sound velocity.

Simple Example

Transmission loss (duct acoustics) calculation - a simple example (one chamber silencer).

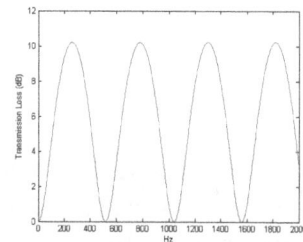

Result of transmission loss (duct acoustics) calculation - a simple example (one chamber silencer). c=520m/s at 400°C; l=0.5m; h=1/3.

Considering we have the most simplest reactive silencer with only one expansion chamber (length l and cross-sectional area S2), with inlet and outlet both having cross-sectional area S1). As we know the transmission matrix of a tube (in this case, the expansion chamber) is,

$$\begin{bmatrix} A & B \\ C & D \end{bmatrix} = \begin{bmatrix} \cos kl & j\dfrac{\rho c}{S_2}\sin kl \\ j\dfrac{S_2}{\rho c}\sin kl & \cos kl \end{bmatrix}.$$

Substitute to the equation of TL above, it can be seen that the TL of this simple reactive silencer is,

$$TL = 10\log_{10}\left(\left|\frac{1}{4}\cos kl + j\frac{S_1}{S_2}\sin kl + j\frac{S_2}{S_1}\sin kl + \cos kl\right|^2\right)$$

$$= 10\log_{10}\left(\cos^2 kl + \frac{1}{4}\left(h + \frac{1}{h}\right)^2\sin^2 kl\right)$$

$$= 10\log_{10}\left(1 + \frac{1}{4}\left(h - \frac{1}{h}\right)^2\sin^2 kl\right),$$

where h is the ratio of the cross-sectional areas and l is the length of the chamber. $k = \omega/c$ is the wave number while c is the sound speed. Note that the transmission loss is zero when l is a multiple of half a wavelength.

As a simple example, consider a one chamber silencer with $h=S_1/S_2=1/3$, at around 400 °C the sound speed is about 520 m/s, with $l=0.5$ m, one easily calculate the TL result shown on the plot on the right. Note that the TL equals zero when frequency is a multiple of $\dfrac{c}{2l}$ and TL peaks when frequency is $\dfrac{c}{4l} + n*\dfrac{c}{2l}$.

Also note that the above calculation is only valid for low-frequency range because at low-frequency range the sound wave can be treated as a plane wave. The TL calculation will start losing its accuracy when the frequency goes above the cutoff frequency, which can be calculated as $f_c = 1.84\dfrac{c}{\pi D}$, where D is diameter of the largest pipe in the structure. In the case above, if for example the muffler body has a diameter of 300mm, then the cut-off frequency is then 1.84*520/pi/0.3=1015 Hz.

Sound Transmission Loss

The sound transmission loss of a partition or a floor are determined by physical factors such as mass and stiffness. In a double layer assembly, such as gypsum wallboard on wood or metal framing, the depth of air spaces, the presence or absence of sound absorbing material, and the degree of mechanical coupling between layers critically affect sound transmission losses and therefore the sound transmission class (STC). An understanding of these factors can lead to improved design and fewer errors. Small changes in the arrangement of materials can yield large changes in STC with little or no increase in cost.

Three interacting physical factors are important in determining whether an occupant of a multifamily dwelling is bothered by noise from neighbours. These are the sound transmission losses of party walls or floors, the level of noise generated in neighbouring homes, and the level of background sound in the occupant's own home. The last two factors can vary widely and one has to select a value of STC that will provide protection for most situations and accept those cases where annoyance is caused by an unusually noisy neighbour or unusually low background sound levels. If the level of background sound is high enough, intruding sounds will be masked and will not be detectable. On the other hand, if the background noise is excessively high, it can interfere with sleep, relaxation, and even conversation. Experience around the world has shown that for occupants of multi-family dwellings to enjoy a reasonable degree of acoustical privacy, the effective STC between dwelling units should be at least 55. Values in excess of STC 60 can be achieved with typical building materials without resorting to extreme designs, although care is needed during the design and construction processes.

Mass Law

The most important physical property controlling the airborne sound transmission loss through an assembly is the mass per unit area of its component layers. The "mass law" is a theoretical rule that applies to most materials in certain frequency ranges. It can be approximated as,

$$TL = 20 \log_{10}(m_s f) - 48$$

where TL is the random incidence transmission loss of the layer; ms is the mass per unit area, kg/m^2; and f is the frequency of the sound wave, Hz. The mass per unit area, ms, is the product of the material density and its thickness.

The mass law equation predicts that each time the frequency of measurement or the mass per unit area of a single layer wall is doubled, the transmission loss increases by about 6 dB. To increase the sound transmission loss of a partition by 12 dB at all frequencies, therefore, the mass per unit area must be increased by a factor of 4; an increase of 18 dB requires an increase by a factor of 8 and so on. Mass per unit area can be increased by increasing thickness or by selecting a more dense material. A single layer of poured concrete 150 mm thick gives an STC of about 55. Layers of this weight are generally the practical limit in normal construction. If a higher STC value than this is necessary, and it often is in high quality construction, it is not economical to continually double the wall or floor thickness to achieve it. Double layer assemblies are a more practical way of getting high STC values without excessive weight.

Effects of Stiffness

Sound transmission losses of partitions and floors are also influenced by stiffness. Transmission loss graphs for stiff materials show dips in particular frequency ranges where the sound transmission losses are reduced below those expected from mass law. This is called the coincidence effect and often leads to a reduced STC rating. Materials with very low stiffness such as sheet lead effectively do not show coincidence dips. Coincidence frequencies for different materials occur in different parts of the acoustical spectrum, sometimes outside the normal range used in building acoustics. The coincidence or critical frequency, f_c, for a given material may be calculated from:

$$f_c = A / t$$

Where A is a constant for each material under consideration and t is the material thickness, mm.

Table gives values of A that permit the calculation of the coincidence frequency for different materials. For example steel has an A value of 12,700 Hz-mm and Equation 2 shows that a 5 mm sheet of steel has a coincidence frequency of 2540 Hz. Similarly, a layer of concrete 100 mm thick would have a coincidence dip at 187 Hz. Figure shows idealized transmission loss curves, including coincidence dips, for some common materials. Materials A, B, and C all have the same mass per unit area but quite different STC ratings because of differing coincidence effects. Because the critical frequencies of concrete and plywood in the thicknesses commonly used fall in the frequency range that is important in building acoustics (100 to 4000 Hz), they are particularly vulnerable to STC reductions due to the effects of coincidence. For gypsum wallboard the coincidence frequency is quite high and the effect on the STC is usually less. The depth of the coincidence dip is determined by the energy losses in the material and at its edges where it is in contact with other materials in the supporting structure. The greater the energy losses, the shallower the coincidence dip and the less the effect on the STC.

Table: Surface Mass, m_s, for 1 mm Thickness and Constant, A, for Calculation of Critical Frequency, f_c, of Some Common Building Materials.

Material	m_s kg/m² per mm	A Hz-mm
Aluminium	2.7	12,900
Concrete, dense poured	2.3	18,700
Hollow concrete block (nominal thickness, 150 mm)	1.1	20,900
Fir timber	0.55	8,900
Glass	2.5	15,200
Lead	11.0	55,900
Plexiglas or Lucite	1.15	30,800
Steel	7.7	12,700
Gypsum board	0.82	39,000
Plywood	0.6	21,700

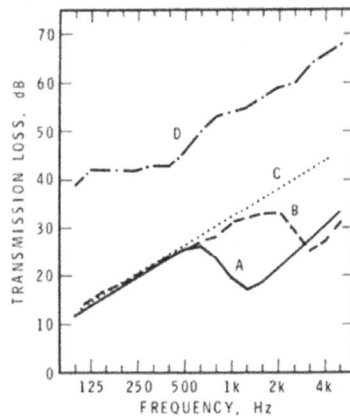

Figure: Transmission losses of typical single-leaf walls, A: 16 mm plywood, 10 kg/m², STC 21; B: 13 mm wallboard, 10 kg/m², STC 28; C: 1.3 mm steel, 10 kg/m², STC 30; D: 100 mm concrete, 235 kg/m², STC 52.

When two layers of material such as wallboard are glued firmly together, they behave like a single thick layer with an associated lowering of the coincidence frequency. If the layers are only held together loosely (with screws for example) so that they can slide over each other to some extent during bending motions, then the coincidence frequency does not move to lower frequencies and the friction between the layers can introduce some extra energy losses.

Double Layer Assemblies

When lightweight construction and high STC values are desired, double layer constructions must be used. These can be very effective but introduce additional effects that must be appreciated if double layer designs are to be successful. Important factors, in addition to the masses of the component layers, are the depth of the air space, the use of sound absorbing materials within the air spaces, and the rigidity of the mechanical coupling between the layers. The ideal double layer assembly has no rigid mechanical connection between its two surfaces.

In a double layer wall or floor the air trapped between the two layers acts as a spring and a resonance, called the mass-air-mass resonance, occurs at a frequency f_{mam} given by:

$$f_{mam} = 1897\sqrt{m_1 + m_2} / \sqrt{Dm_1m_2}$$

Where m_1, m_2 are the surface masses of the layers, kg/m²; and D is the distance between the layers, mm. The larger the air space or the heavier the materials, the lower the frequency at which resonance occurs. At frequencies below the resonance frequency, the layers are coupled by the air in the cavity and the TL is that due to the sum of their masses. Close to the resonance frequency, however, the transmission losses are usually lower than this. Above the resonance frequency, the sound transmission loss increases much more rapidly than mass law predictions for the sum of the masses. Figure gives an example of the benefits to be obtained by increasing the air space between the two layers of a wall. Generally, partitions should be designed so that the mass-air-mass resonance is below 80 Hz.

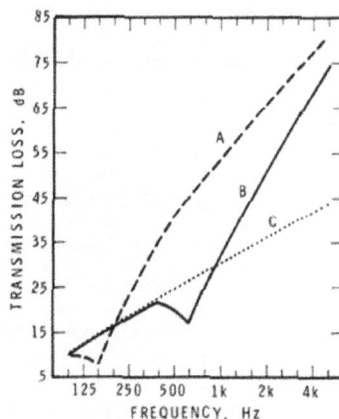

In figure Effect of air space on ideal double walls with 0.5 mm steel on each face, sound absorbing material in the cavity and no rigid mechanical connections between the faces. A has an airspace of 100 mm, a resonance dip at 135 Hz, and an STC of 29; B has an airspace of 5 mm, a resonance dip at 630 Hz, and an STC of 24. Curve C represents mass law predictions for a single 1 mm steel sheet and has an STC of 28.

Standing wave resonances between the layers of a double layer wall or floor occur at relatively high frequencies and the sound transmission losses can be further reduced by them. The negative effects of most of these resonances can be reduced by the addition of sound absorbing material inside the cavities. For normal wall thicknesses (around 100 mm) the density and the thickness of the sound absorbing material is not a very important factor. Increasing the thickness beyond about 75 mm has little effect on the STC rating, although, for floors or walls that are significantly thicker than normal, it becomes more important to use thicker layers of glass fibre. The type of glass fibre or mineral wool insulation normally used for thermal purposes absorbs sound well and is quite adequate for use inside double layer walls as a sound absorbing material.

Gypsum Board Walls

The mechanical connection between the layers of wallboard can be reduced by the use of staggered wood studs, separate rows of wood studs, or a single row of wood studs with resilient metal furring strips to support the wallboard layers independently of each other. Nonload-bearing steel studs are usually resilient enough to provide adequate mechanical decoupling between the layers. Good results have also been obtained using 150 mm loadbearing steel studs in conjunction with resilient channels. Table gives some representative STC values for typical constructions. The presence of the sound absorbing material increases the STC by about 8 points relative to the same wall without sound absorbing material. The thickness of the wallboard is not specified in the table since it is only a guide. Walls with 16 mm board would be better than those with 13 mm board by a few points. The table shows that STC values of 60 or more can be obtained if the air space is large enough and enough wallboard is used. Such values have been measured in buildings as well as in laboratories.

Table: STC Ratings for Walls Formed From Two Layers of Wallboard*.

Wall construction	Number of Layers of Wallboard on Each Wall Surface		
	1 + 1	1 + 2	2 + 2
38 x 89 mm wood studs with resilient steel channels on one side	48 [40]	52 [44]	56 [52]
Staggered 38 x 89 mm wood studs	50 [41]	53 [47]	55 [52]
Double row of 38 x 89 mm wood studs with small gap between them	57 [46]	60 [52]	63 [57]
90 mm steel studs	45 [39]	49 [45]	56 [50]
150 mm load-bearing steel studs with resilient metal channels on one side	58	60	63
*Values not in brackets are for walls filled with sound absorbing material. Values in brackets are for walls without sound absorbing material.			

In contrast to the values in the table, the common internal partition used in single family homes with drywall attached directly to both sides of the wood studs has an STC rating of about 33. The addition of sound absorbing material in this wall increases its rating by about only 3 points, because the sound energy is transmitted directly from one layer of wallboard to the other through the studs. The sound absorbing material in the cavity is of much less benefit than it would be if the layers were decoupled, in which case most of the sound would be transmitted through the air in the cavity. Rigid mechanical connections are the acoustical equivalent of an electrical short circuit or a thermal bridge in an insulated wall and should be avoided.

Concrete Block Walls

Concrete block walls commonly have wallboard applied to each face as a finishing material. Resilient connections and sound absorbing material in cavities are as important in block walls as they are in wood or steel frame construction. The mass-air-mass resonance is also important. If the air gap behind the wallboard is too small, the sound transmission losses can be reduced relative to the unfinished wall. For a single layer of wallboard attached to a concrete block wall, the air space should be greater than 60 mm to meet the 80 Hz criterion previously noted. For a double layer of wallboard, the space may be as small as 35 mm. These air spaces are larger than those typically used in concrete block walls where relatively thin wood furring strips are often used to attach the wallboard. Even using adhesive to attach wallboard directly to concrete can result in a thin film of air a few mm in thickness trapped behind the wallboard and a deleterious mass-air-mass resonance. Reduced sound transmission losses caused by the mass-air-mass resonance are often the cause of a low STC for a potentially good concrete block wall. An increase in the air gap of just a few centimetres can increase the STC considerably.

Some concrete blocks are slightly porous so that the effective thickness of the air layer behind the wallboard is greater than its physical dimensions and the mass-air-mass resonance is lower than expected. This characteristic can only be verified by acoustical testing, however. Sound transmission class ratings for some concrete block constructions can be found in references 2, 3 and 4. STC ratings of 60 or more can readily be achieved with a concrete block wall if it is correctly designed and constructed.

Flanking Transmission

In laboratory measurements of airborne sound transmission, the only significant sound transmission path between the test rooms is through the test partition or the test floor itself. In real buildings, however, sound travels between suites indirectly by way of the surrounding constructions as well as directly through the common wall or floor assembly. These less obvious paths for the sound are called flanking paths and, in a poor design, they can transmit more sound energy than the direct path through the common wall or floor. All of these paths comprise a system that must be considered as a whole so that assemblies built in the field can attain values close to those in laboratory tests. Floors are particularly prone to increased impact sound transmission because of flanking transmission through the supporting structure. Flanking transmission is beyond the scope of this Digest, however. If the physical factors that control the STC of partitions and floor assemblies are understood, the principles can be applied to all transmission paths.

9

Acoustic Measurement

Acoustic measurement refers to the measurement of the values which describe sound in terms of their intensities as well as qualitative features. Some of its applications include acoustic microscopy, beamforming, spectrograms, and acoustic interferometers. This chapter closely examines these concepts and applications of acoustic measurement to provide an extensive understanding of the subject.

Acoustic Measurements are the measurements of the values that describe sounds and noises in terms of their intensities and various qualitative features (such as their spectra or the growth and decay of the sound over time). The principal values measured in acoustics are sound pressure, sound intensity, vibration velocity and particle displacement, the frequency and period of vibrations, propagation velocity, damping factor, and others. The most important characteristic is sound pressure, because the human ear responds to the sound wave pressure.

Acoustic measurements are closely linked with electrical measurements and are performed chiefly with electronic measuring instruments. The difficulties experienced in acoustic measurements are due to the complicated spatial distribution of the acoustic values in rooms as well as the variability of sounds and noises over time.

For sound pressure measurements a standard microphone is used in air or a hydrophone in water. The receiving portion of these instruments transforms the received acoustic signals (pressures) into proportional electric voltages, which are then fed to the input of amplifiers with indicators for readout. A sound-level meter is used for measuring various noises.

An important part of acoustic measurements is building and architectural acoustics—the measurements of the sound isolation of partitions and coverings and of the sound absorption factor of different structural coverings (such as plaster, upholstery, and floors).

There are other kinds of acoustic measurements such as measurements of waveguide characteristics, tests of acoustic instruments for communication and broadcasting (that is, of acoustic transmitters and receivers), tests of tape recorders and record players, and of telephone communication. Of special importance is the category of subjective measurements of the hearing sensitivity of people and its deviations from the norm (audiometry).

Acoustic Microscopy

Acoustic microscopy is microscopy that employs very high or ultra high frequency ultrasound. Acoustic microscopes operate non-destructively and penetrate most solid materials to make visible images of internal features, including defects such as cracks, delaminations and voids.

Types of Acoustic Microscopes

In the half-century since the first experiments directly leading to the development of acoustic microscopes, at least three basic types of acoustic microscope have been developed. These are the scanning acoustic microscope (SAM), confocal scanning acoustic microscope (CSAM), and C-mode scanning acoustic microscope (C-SAM).

More recently acoustic microscopes based around picosecond ultrasonics systems have demonstrated acoustic imaging in cells using sub-optical wavelengths working with ultrasonic frequencies into the multi-GHz. Since the vast majority of acoustic microscopes in use today are C-SAM type instruments, this discussion will be limited to these instruments.

Behavior of Ultrasound in Materials

Ultrasound is broadly defined as any sound having a frequency above 20 kHz, which is approximately the highest frequency that can be detected by the human ear. However, the acoustic microscopes emit ultrasound ranging from 5 MHz to beyond 400 MHz so that micrometre size resolution can be achieved. The ultrasound that penetrates a sample may be scattered, absorbed or reflected by the internal features or the material itself. These actions are analogous to the behavior of light. Ultrasound that is reflected from an internal feature, or (in some applications) that has traveled through the entire thickness of the sample, is used to make acoustic images.

Sample Types and Preparation

Samples need no special treatment before acoustic imaging, but they should be able to withstand at least brief exposure to water or to another fluid, since air is a very poor transmitter of high frequency acoustic energy from the transducer. The sample may be completely immersed in the water, or scanned with a narrow stream of water. Alternately, alcohols and other fluids can be used so as to not contaminate the sample. Samples typically have at least one flat surface that can be scanned, although cylindrical and spherical samples can also be scanned with the proper fixtures. In the following paragraphs, the sample being described is a plastic-encapsulated integrated circuit.

Ultrasonic Frequencies

The ultrasonic frequencies pulsed into samples by the transducers of acoustic microscopes range from a low of 10 MHz (rarely, 5 MHz) to a high of 400 MHz or more. Across this spectrum of frequencies there is a trade-off of penetration and resolution. Ultrasound at low frequencies such as 10 MHz penetrates deeper into materials than ultrasound at higher frequencies, but the spatial resolution of the acoustic image is less. On the other hand, ultrasound at very high frequencies do not penetrate deeply, but provide acoustic images having very high resolution. The frequency chosen to image a particular sample will depend on the geometry of the part and on the materials involved.

The acoustic image of the plastic-encapsulated IC below was made using a 30 MHz transducer because this frequency provides a good compromise between penetration and image resolution.

Scanning Process

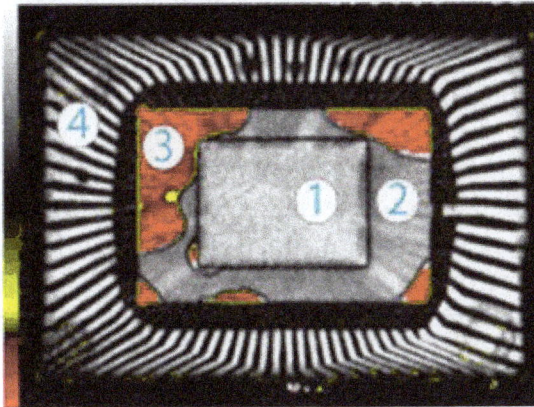

In the acoustic image ultrasound was pulsed through the black mold compound (plastic), and reflected from the interface between the overlying mold compound and the top surface of the silicon die, the top surface of the die paddle, delaminations (red) on top of the die paddle, and the outer portion (lead fingers) of the lead frame.

The ultrasonic transducer raster-scans the top surface of the sample. Several thousand pulses enter the sample each second. Each pulse may be scattered or absorbed in passing through homogeneous parts of the sample. At material interfaces, a portion of the pulse is reflected back to the transducer, where it is received and its amplitude recorded.

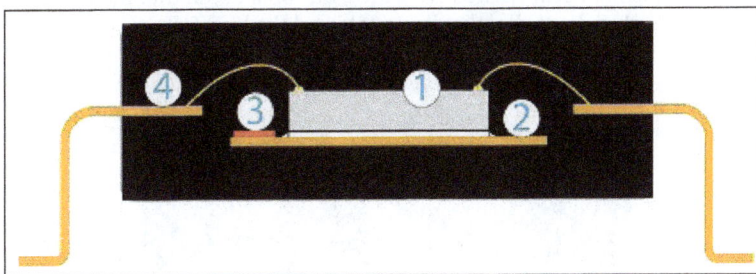

Side-view diagram

The portion of the pulse that is reflected is determined by the acoustic impedance, Z, of the each material that meets at the interface. The acoustic impedance of a given material is the material's density multiplied by the speed of ultrasound in that material. When a pulse of ultrasound encounters an interface between two materials, the degree of ultrasonic reflection from that interface is governed by this formula:

$$R = \frac{(z_2 - z_1)}{(z_2 + z_1)}$$

where R is the fraction of reflection, and z_1 and z_2 are the acoustic impedances of the two materials, analogous to refractive index in light propagation.

If both materials are typical solids, the degree of reflection will be moderate, and a significant portion of the pulse will travel deeper into the sample, where it may be in part reflected by deeper material interfaces. If one of the materials is a gas such as air – as in the case with delaminations, cracks and voids – the degree of reflection at the solid-to-gas interface is near 100%, the amplitude of the reflected pulse is very high, and practically none of the pulse travels deeper into the sample.

Gating of the Return Echoes

A pulse of ultrasound from the transducer travel nanoseconds or microseconds to reach an internal interface and are reflected back to the transducer. If there are several internal interfaces at different depths, the echoes arrive at the transducer at different times. Planar acoustic images do not often use all return echoes from all depths to make the visible acoustic image. Instead, a time window is created that accepts only those return echoes from the depth of interest. This process is known as "gating" the return echoes.

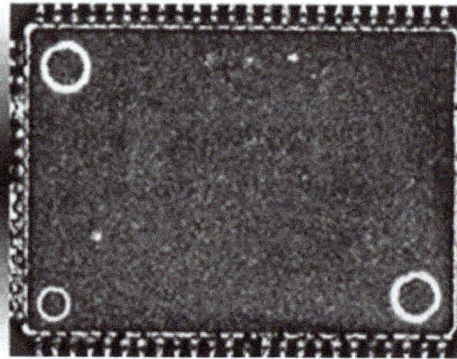

In the plastic-encapsulated IC, gating was on a depth that included the silicon die, the die paddle and the lead frame.

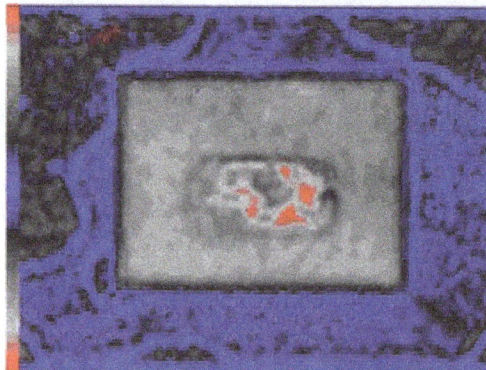

Still scanning the top of the sample, the gating of the return echoes was then changed to include only the plastic encapsulant (mold compound) above the die. The resulting acoustic image is shown above. It shows the structure of the particle-filled plastic mold compound, as well as the circular mold marks at the top surface of the component. The small white features are voids (trapped bubbles) in the mold compound. (These voids are also visible in the previous image as dark acoustic shadows).

Gating was then changed to include only depth of the die attach material that attaches the silicon die to the die paddle. The die, the die paddle, and other features above and below the die attach depth are ignored. In the resulting acoustic, shown above slightly magnified, the red areas are voids (defects) in the die attach material.

Finally, the plastic-encapsulated IC was flipped over and imaged from the back side. The return echoes were gated on the depth where the backside mold compound interfaces with the back side of the die paddle. The small black dots in the acoustic image above are small voids (trapped bubbles) in the mold compound.

Other Image Types

The acoustic images shown above are all planar images, so named because they make visible a horizontal plane within the sample. The acoustic data received in the return echo signals can also be used to make other types of images, including three-dimensional images, cross-sectional images, and thru-scan images.

Range of Applications

The samples imaged by acoustic microscopes are typically assemblies of one or more solid materials that have at least one surface that is either flat or regularly curved. The depth of interest may involve an internal bond between materials, or a depth at which a defect may occur in a homogeneous material. In addition, samples may be characterized without imaging to determine, e.g., their acoustic impedance.

Because of their ability to find visualize features non-destructively, acoustic microscopes are widely used in the production of electronic components and assemblies for quality control, reliability and failure analysis. Usually the interest is in finding and analyzing internal defects such as delaminations, cracks and voids, although an acoustic microscope may also be used simply to verify (by material characterization or imaging, or both) that a given part or a given material meets specifications or, in some instances, is not counterfeit. Acoustic microscopes are also used to image printed circuit boards and other assemblies.

There are in addition numerous applications outside of electronics. The assembly of numerous medical products uses acoustic microscopes to investigate internal bonds and features. For example, a polymer film may be imaged to examine its bond to a multi-channel plastic plate used in blood analysis. In many industries, products that involve tubing, ceramic materials, composite materials or some types of welds may be imaged acoustically.

A more recent application is the use of acoustic microscopy to the diagnosis of the paint layers of painted art and other objects.

Beamforming

Beamforming or spatial filtering is a signal processing technique used in sensor arrays for directional signal transmission or reception. This is achieved by combining elements in an antenna array in such a way that signals at particular angles experience constructive interference while others experience destructive interference. Beamforming can be used at both the transmitting and receiving ends in order to achieve spatial selectivity. The improvement compared with omnidirectional reception/transmission is known as the directivity of the array.

Beamforming can be used for radio or sound waves. It has found numerous applications in radar, sonar, seismology, wireless communications, radio astronomy, acoustics and biomedicine. Adaptive beamforming is used to detect and estimate the signal of interest at the output of a sensor array by means of optimal (e.g. least-squares) spatial filtering and interference rejection.

Techniques

To change the directionality of the array when transmitting, a beamformer controls the phase and relative amplitude of the signal at each transmitter, in order to create a pattern of constructive and destructive interference in the wavefront. When receiving, information from different sensors is combined in a way where the expected pattern of radiation is preferentially observed.

For example, in sonar, to send a sharp pulse of underwater sound towards a ship in the distance, simply simultaneously transmitting that sharp pulse from every sonar projector in an array fails because the ship will first hear the pulse from the speaker that happens to be nearest the ship, then later pulses from speakers that happen to be further from the ship. The beamforming technique involves sending the pulse from each projector at slightly different times (the projector closest to the ship last), so that every pulse hits the ship at exactly the same time, producing the effect of a single strong pulse from a single powerful projector. The same technique can be carried out in air using loudspeakers, or in radar/radio using antennas.

In passive sonar, and in reception in active sonar, the beamforming technique involves combining delayed signals from each hydrophone at slightly different times (the hydrophone closest to the target will be combined after the longest delay), so that every signal reaches the output at exactly the same time, making one loud signal, as if the signal came from a single, very sensitive hydrophone. Receive beamforming can also be used with microphones or radar antennas.

With narrow-band systems the time delay is equivalent to a "phase shift", so in this case the array of antennas, each one shifted a slightly different amount, is called a phased array. A narrow band system, typical of radars, is one where the bandwidth is only a small fraction of the center frequency. With wide band systems this approximation no longer holds, which is typical in sonars.

In the receive beamformer the signal from each antenna may be amplified by a different "weight." Different weighting patterns (e.g., Dolph-Chebyshev) can be used to achieve the desired sensitivity patterns. A main lobe is produced together with nulls and sidelobes. As well as controlling the main lobe width (beamwidth) and the sidelobe levels, the position of a null can be controlled. This is useful to ignore noise or jammers in one particular direction, while listening for events in other directions. A similar result can be obtained on transmission.

Beamforming techniques can be broadly divided into two categories:

- Conventional (fixed or switched beam) beamformers.

- Adaptive beamformers or phased array:

  ○ Desired signal maximization mode.

  ○ Interference signal minimization or cancellation mode.

Conventional beamformers, such as the Butler matrix, use a fixed set of weightings and time-delays (or phasings) to combine the signals from the sensors in the array, primarily using only information about the location of the sensors in space and the wave directions of interest. In contrast, adaptive beamforming techniques (e.g., MUSIC, SAMV) generally combine this information with properties of the signals actually received by the array, typically to improve rejection of unwanted signals from other directions. This process may be carried out in either the time or the frequency domain.

As the name indicates, an adaptive beamformer is able to automatically adapt its response to different situations. Some criterion has to be set up to allow the adaptation to proceed such as minimizing the total noise output. Because of the variation of noise with frequency, in wide band systems it may be desirable to carry out the process in the frequency domain.

Beamforming can be computationally intensive. Sonar phased array has a data rate low enough that it can be processed in real-time in software, which is flexible enough to transmit or receive in several directions at once. In contrast, radar phased array has a data rate so high that it usually requires dedicated hardware processing, which is hard-wired to transmit or receive in only one direction at a time. However, newer field programmable gate arrays are fast enough to handle radar data in real-time, and can be quickly re-programmed like software, blurring the hardware/software distinction.

Sonar Beamforming Requirements

Sonar beamforming utilizes a similar technique to electromagnetic beamforming, but varies considerably in implementation details. Sonar applications vary from 1 Hz to as high as 2 MHz, and array elements may be few and large, or number in the hundreds yet very small. This will shift sonar beamforming design efforts significantly between demands of such system components as the "front end" (transducers, pre-amplifiers and digitizers) and the actual beamformer computational hardware downstream. High frequency, focused beam, multi-element imaging-search sonars and acoustic cameras often implement fifth-order spatial processing that places strains equivalent to Aegis radar demands on the processors.

Many sonar systems, such as on torpedoes, are made up of arrays of up to 100 elements that must accomplish beam steering over a 100 degree field of view and work in both active and passive modes.

Sonar arrays are used both actively and passively in 1-, 2-, and 3-dimensional arrays.

- 1-dimensional "line" arrays are usually in multi-element passive systems towed behind ships and in single- or multi-element side-scan sonar.

- 2-dimensional "planar" arrays are common in active/passive ship hull mounted sonars and some side-scan sonar.

- 3-dimensional spherical and cylindrical arrays are used in 'sonar domes' in the modern submarine and ships.

Sonar differs from radar in that in some applications such as wide-area-search all directions often need to be listened to, and in some applications broadcast to, simultaneously. Thus a multibeam system is needed. In a narrowband sonar receiver the phases for each beam can be manipulated entirely by signal processing software, as compared to present radar systems that use hardware to 'listen' in a single direction at a time.

Sonar also uses beamforming to compensate for the significant problem of the slower propagation speed of sound as compared to that of electromagnetic radiation. In side-look-sonars, the speed of the towing system or vehicle carrying the sonar is moving at sufficient speed to move the sonar out of the field of the returning sound "ping". In addition to focusing algorithms intended to improve reception, many side scan sonars also employ beam steering to look forward and backward to "catch" incoming pulses that would have been missed by a single sidelooking beam.

Digital, Analog and Hybrid

For receive (but not transmit), there is a distinction between analog and digital beamforming. For example, if there are 100 sensor elements, the "digital beamforming" approach entails that each of the 100 signals passes through an analog-to-digital converter to create 100 digital data streams. Then these data streams are added up digitally, with appropriate scale-factors or phase-shifts, to get the composite signals. By contrast, the "analog beamforming" approach entails taking the 100 analog signals, scaling or phase-shifting them using analog methods, summing them, and then usually digitizing the single output data stream.

Digital beamforming has the advantage that the digital data streams (100 in this example) can be manipulated and combined in many possible ways in parallel, to get many different output signals in parallel. The signals from every direction can be measured simultaneously, and the signals can be integrated for a longer time when studying far-off objects and simultaneously integrated for a shorter time to study fast-moving close objects, and so on. This cannot be done as effectively for analog beamforming, not only because each parallel signal combination requires its own circuitry, but more fundamentally because digital data can be copied perfectly but analog data cannot. (There is only so much analog power available, and amplification adds noise.) Therefore, if the received analog signal is split up and sent into a large number of different signal combination circuits, it can reduce the signal-to-noise ratio of each.

In MIMO communication systems with large number of antennas, so called massive MIMO systems, the beamforming algorithms executed at the digital baseband can get very complex. In addition, if all beamforming is done at baseband, each antenna needs its own RF feed. At high frequencies and with large number of antenna elements, this can be very costly, and increase loss and complexity in the system. To remedy these issues, hybrid beamforming has been suggested where some of the beamforming is done using analog components and not digital.

There are many possible different functions that can be performed using analog components instead of at the digital baseband.

For Speech Audio

Beamforming can be used to try to extract sound sources in a room, such as multiple speakers in the cocktail party problem. This requires the locations of the speakers to be known in advance, for example by using the time of arrival from the sources to mics in the array, and inferring the locations from the distances.

Compared to carrier-wave telecommunications, natural audio contains a variety of frequencies. It is advantageous to separate frequency bands prior to beamforming because different frequencies have different optimal beamform filters (and hence can be treated as separate problems, in parallel, and then recombined afterward). Properly isolating these bands involves specialized non-standard filter banks. In contrast, for example, the standard fast Fourier transform (FFT) band-filters implicitly assume that the only frequencies present in the signal are exact harmonics; frequencies which lie between these harmonics will typically activate all of the FFT channels (which is not what is wanted in a beamform analysis). Instead, filters can be designed in which only local frequencies are detected by each channel (while retaining the recombination property to be able to reconstruct the original signal), and these are typically non-orthogonal unlike the FFT basis.

Spectrogram

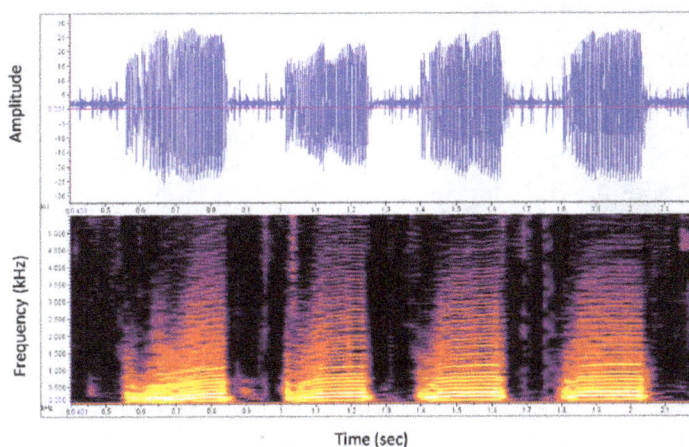

Spectrogram of the spoken words "nineteenth century". Frequencies are shown increasing up the vertical axis, and time on the horizontal axis. The legend to the right shows that the color intensity increases with the density.

A spectrogram is a visual representation of the spectrum of frequencies of a signal as it varies with time. When applied to an audio signal, spectrograms are sometimes called sonographs, voiceprints, or voicegrams. When the data is represented in a 3D plot they may be called waterfalls.

Spectrograms are used extensively in the fields of music, sonar, radar, and speech processing, seismology, and others. Spectrograms of audio can be used to identify spoken words phonetically, and to analyse the various calls of animals.

A spectrogram can be generated by an optical spectrometer, a bank of band-pass filters, by Fourier transform or by a wavelet transform (in which case it is also known as a scaleogram).

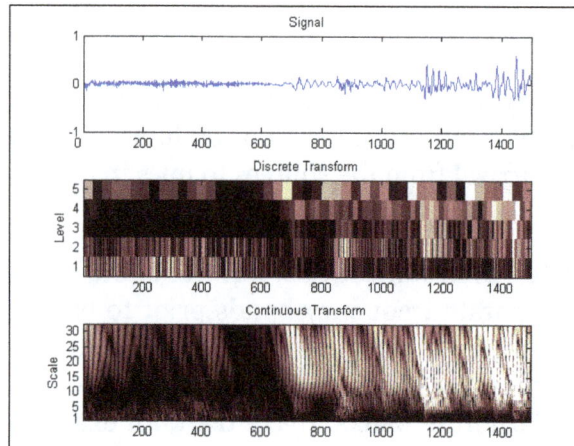

Scaleograms from the DWT and CWT for an audio sample

A spectrogram is usually depicted as a heat map, i.e., as an image with the intensity shown by varying the colour or brightness.

Format

A common format is a graph with two geometric dimensions: One axis represents time, and the other axis represents frequency; a third dimension indicating the amplitude of a particular frequency at a particular time is represented by the intensity or color of each point in the image.

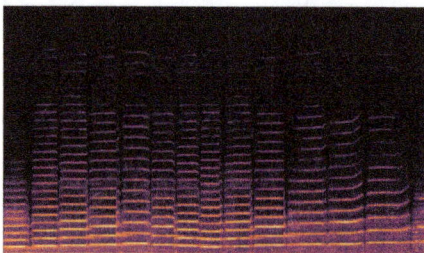

Spectrogram of this recording of a violin playing. Note the harmonics occurring at whole-number multiples of the fundamental frequency.

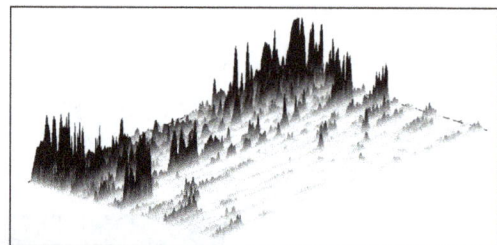

3D surface spectrogram of a part from a music piece.

There are many variations of format: Sometimes the vertical and horizontal axes are switched, so time runs up and down; sometimes as a waterfall plot where the amplitude is represented by height of a 3D surface instead of color or intensity. The frequency and amplitude axes can be either linear or logarithmic, depending on what the graph is being used for. Audio would usually be represented with a logarithmic amplitude axis (probably in decibels, or dB), and frequency would be linear to emphasize harmonic relationships, or logarithmic to emphasize musical, tonal relationships.

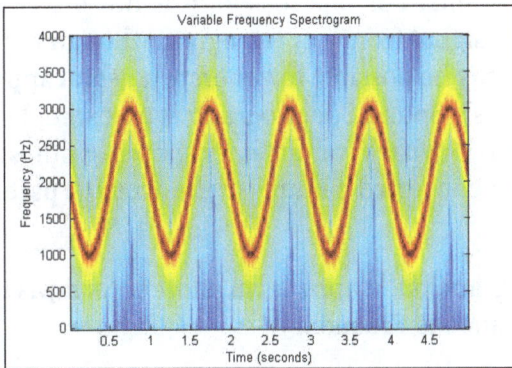

Spectrogram of an FM signal. In this case the signal frequency is modulated with a sinusoidal frequency vs. time profile.

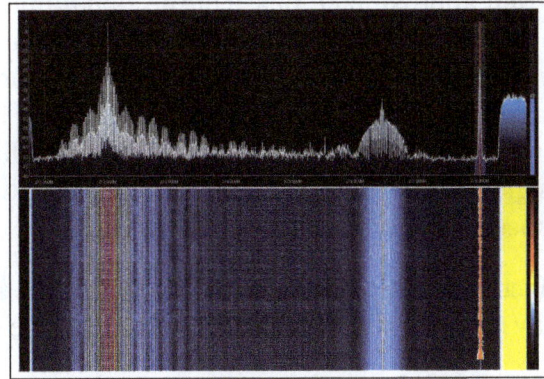

Spectrum above and waterfall (Spectrogram) below of an 8MHz wide PAL-I Television signal.

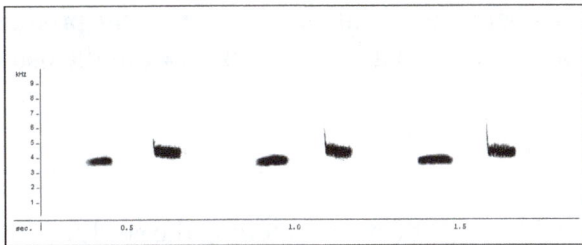

Spectrogram of great tit song.

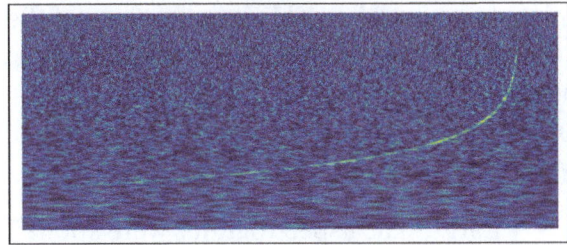

Spectrogram of a gravitational wave (GW170817).

Generation

Spectrograms of light may be created directly using an optical spectrometer over time. Spectrograms may be created from a time-domain signal in one of two ways: approximated as a filterbank that results from a series of band-pass filters (this was the only way before the advent of modern digital signal processing), or calculated from the time signal using the Fourier transform. These two methods actually form two different time–frequency representations, but are equivalent under some conditions.

The bandpass filters method usually uses analog processing to divide the input signal into frequency bands; the magnitude of each filter's output controls a transducer that records the spectrogram as an image on paper.

Creating a spectrogram using the FFT is a digital process. Digitally sampled data, in the time domain, is broken up into chunks, which usually overlap, and Fourier transformed to calculate the magnitude of the frequency spectrum for each chunk. Each chunk then corresponds to a vertical line in the image; a measurement of magnitude versus frequency for a specific moment in time (the midpoint of the chunk). These spectrums or time plots are then "laid side by side" to form the image or a three-dimensional surface, or slightly overlapped in various ways, i.e. windowing. This process essentially corresponds to computing the squared magnitude of the short-time Fourier transform (STFT) of the signal $s(t)$ — that is, for a window width ω, spectrogram$(t,\omega) = |\text{STFT}(t,\omega)|^2$.

Limitations and Resynthesis

From the formula above, it appears that a spectrogram contains no information about the exact, or even approximate, phase of the signal that it represents. For this reason, it is not possible to reverse the process and generate a copy of the original signal from a spectrogram, though in situations where the exact initial phase is unimportant it may be possible to generate a useful approximation of the original signal. The Analysis & Resynthesis Sound Spectrograph is an example of a computer program that attempts to do this. The Pattern Playback was an early speech synthesizer, designed at Haskins Laboratories in the late 1940s, that converted pictures of the acoustic patterns of speech (spectrograms) back into sound.

In fact, there is some phase information in the spectrogram, but it appears in another form, as time delay (or group delay) which is the dual of the Instantaneous Frequency.

The size and shape of the analysis window can be varied. A smaller (shorter) window will produce more accurate results in timing, at the expense of precision of frequency representation. A larger (longer) window will provide a more precise frequency representation, at the expense of precision in timing representation. This is an instance of the Heisenberg uncertainty principle, that the product of the precision in two conjugate variables is less than or equal to a constant (B*T>=1 in the usual notation).

Applications

- Early analog spectrograms were applied to a wide range of areas including the study of bird calls (such as that of the great tit), with current research continuing using modern digital equipment and applied to all animal sounds. Contemporary use of the digital spectrogram is especially useful for studying frequency modulation (FM) in animal calls. Specifically, the distinguishing characteristics of FM chirps, broadband clicks, and social harmonizing are most easily visualized with the spectrogram.

- Spectrograms are useful in assisting in overcoming speech deficits and in speech training for the portion of the population that is profoundly deaf.

- The studies of phonetics and speech synthesis are often facilitated through the use of spectrograms.

- By reversing the process of producing a spectrogram, it is possible to create a signal whose spectrogram is an arbitrary image. This technique can be used to hide a picture in a piece of audio and has been employed by several electronic music artists.

- Some modern music is created using spectrograms as an intermediate medium; changing the intensity of different frequencies over time, or even creating new ones, by drawing them and then inverse transforming.

- Spectrograms can be used to analyze the results of passing a test signal through a signal processor such as a filter in order to check its performance.

- High definition spectrograms are used in the development of RF and microwave systems.

- Spectrograms are now used to display scattering parameters measured with vector network analyzers.

- The US Geological Survey now provides real-time spectrogram displays from seismic stations.

- Spectrograms can be used with recurrent neural networks for speech recognition.

Acoustic Interferometer

Acoustic interferometer is a device for measuring the velocity and absorption of sound waves in a gas or liquid. A vibrating crystal creates the waves that are radiated continuously into the fluid medium, striking a movable reflector placed accurately parallel to the crystal source. The waves are then reflected back to the source. The strength of the standing wave pattern set up between the source and the reflector as the distance between source and reflector is varied, or as the frequency is varied, indicates the absorption by the medium. The velocity at which the waves travel may be determined from the distance between the peaks in the pattern of standing waves. This method has been brought to a high degree of accuracy and has been used for determining the characteristics of seawater for sonar applications.

References

- Geier, Eric. "All about beamforming, the faster Wi-Fi you didn't know you needed". PC World. IDG Consumer & SMB. Retrieved 19 October 2015

- Acoustic-Measurements: encyclopedia2.thefreedictionary.com, Retrieved 30 June, 2019

- Kessler, L.W.; Yuhas, D.E. (1979). "Acoustic microscopy—1979". Proceedings of the IEEE. 67 (4): 526. Doi:10.1109/PROC.1979.11281

- Sejdic, E.; Djurovic, I.; Stankovic, L. (August 2008). "Quantitative Performance Analysis of Scalogram as Instantaneous Frequency Estimator". IEEE Transactions on Signal Processing. 56 (8): 3837–3845. doi:10.1109/TSP.2008.924856. ISSN 1053-587X

- Acoustic-interferometer, technology: britannica.com, Retrieved 21 April, 2019

Index

www.ingramcontent.com/pod-product-compliance
Lightning Source LLC
Chambersburg PA
CBHW082046190326
41458CB00010B/3475